A Life Cycle for Clusters?

Contributions to Economics

www.springer.com/series/1262

Further volumes of this series can
be found at our homepage.

Pablo Coto-Millán
General Equilibrium and Welfare
2002. ISBN 7908-1491-1

Wojciech W. Charemza/Krystyna Strzala (Eds.)
East European Transition and EU
Enlargement
2002. ISBN 3-7908-1501-1

Natalja von Westernhagen
Systemic Transformation, Trade and
Economic Growth
2002. ISBN 3-7908-1521-7

Josef Falkinger
A Theory of Employment in Firms
2002. ISBN 3-7908-1520-9

Engelbert Plassmann
Econometric Modelling of European Money
Demand
2003. ISBN 3-7908-1522-5

Reginald Loyen/Erik Buyst/Greta Devos (Eds.)
Struggeling for Leadership:
Antwerp-Rotterdam Port Competition
between 1870–2000
2003. ISBN 3-7908-1524-1

Pablo Coto-Millán
Utility and Production, 2nd Edition
2003. ISBN 3-7908-1423-7

Emilio Colombo/John Driffill (Eds.)
The Role of Financial Markets
in the Transition Process
2003. ISBN 3-7908-0004-X

Guido S. Merzoni
Strategic Delegation in Firms
and in the Trade Union
2003. ISBN 3-7908-1432-6

Jan B. Kuné
On Global Aging
2003. ISBN 3-7908-0030-9

Sugata Marjit, Rajat Acharyya
International Trade, Wage Inequality
and the Developing Economy
2003. ISBN 3-7908-0031-7

Francesco C. Billari/Alexia Prskawetz (Eds.)
Agent-Based Computational Demography
2003. ISBN 3-7908-1550-0

Georg Bol/Gholamreza Nakhaeizadeh/
Svetlozar T. Rachev/Thomas Ridder/
Karl-Heinz Vollmer (Eds.)
Credit Risk
2003. ISBN 3-7908-0054-6

Christian Müller
Money Demand in Europe
2003. ISBN 3-7908-0064-3

Cristina Nardi Spiller
The Dynamics of the Price Structure
and the Business Cycle
2003. ISBN 3-7908-0063-5

Michael Bräuninger
Public Debt and Endogenous Growth
2003. ISBN 3-7908-0056-1

Brigitte Preissl/Laura Solimene
The Dynamics of Clusters and Innovation
2003. ISBN 3-7908-0077-5

Markus Gangl
Unemployment Dynamics in the
United States and West Germany
2003. ISBN 3-7908-1533-0

Pablo Coto-Millán (Ed.)
Essays on Microeconomics
and Industrial Organisation, 2nd Edition
2004. ISBN 3-7908-0104-6

Wendelin Schnedler
The Value of Signals
in Hidden Action Models
2004. ISBN 3-7908-0173-9

Carsten Schröder
Variable Income Equivalence Scales
2004. ISBN 3-7908-0183-6

Wilhelm J. Meester
Locational Preferences of Entrepreneurs
2004. ISBN 3-7908-0178-X

Russel Cooper/Gary Madden (Eds.)
Frontiers of Broadband, Electronic and
Mobile Commerce
2004. ISBN 3-7908-0087-2

Sardar M.N. Islam
Empirical Finance
2004. ISBN 3-7908-1551-9

Jan-Egbert Sturm/Timo Wollmershäuser (Eds.)
Ifo Survey Data in Business Cycle
and Monetary Policy Analysis
2005. ISBN 3-7908-0174-7

Bernard Michael Gilroy/Thomas Gries/
Willem A. Naudé (Eds.)
Multinational Enterprises, Foreign Direct
Investment and Growth in Africa
2005. ISBN 3-7908-0276-X

Günter S. Heiduk/Kar-yiu Wong (Eds.)
WTO and World Trade
2005. ISBN 3-7908-1579-9

Emilio Colombo/Luca Stanca
Financial Market Imperfections
and Corporate Decisions
2006. ISBN 3-7908-1581-0

Birgit Mattil
Pension Systems
2006. ISBN 3-7908-1675-2

Kerstin Press

A Life Cycle for Clusters?

The Dynamics of Agglomeration, Change, and Adaption

With 40 Figures and 16 Tables

Physica-Verlag

A Springer Company

Series Editors
Werner A. Müller
Martina Bihn

Author
Dr. Kerstin Press
Chair for Systems Design
ETH Zürich, D-MTEC
KPL F 29.1
Kreuzplatz 5
8032 Zürich
Switzerland
kerstinpress@web.de

ISSN 1431-1933
ISBN-10   3-7908-1710-4 Physica-Verlag Heidelberg New York
ISBN-13   978-3-7908-1710-2 Physica-Verlag Heidelberg New York

Physica-Verlag is a part of Springer Science+Business Media

springer.com

© Physica-Verlag Heidelberg 2006
Printed in Germany

Cover-Design: Erich Kirchner, Heidelberg

SPIN 11685371   88/3153-5 4 3 2 1 0 – Printed on acid-free and non-aging paper

It is not the strongest of the species that survive, nor the most intelligent, but the one most responsive to change.

Charles Darwin, 1859

Foreword

The current research agenda in economics is witnessing the rediscovery of a set of theories investigating the determinants and effects of the location of economic activity in space. Based on the linkages between supply, production, sales and space in (traditional) location choice theory, reasoning is being extended to explain the emergence of spatial concentrations of economic activity as well as attempts to incorporate the issue of space into those general equilibrium models that still dominate most of mainstream economic research (e.g. the link between space and trade flows analysed in the New Economic Geography).

This renaissance of theoretical and empirical work on the spatial aspects associated with economics and especially the uneven spatial distribution of economic activity is of key interest in the age of globalisation. Increased international mobility of goods and production factors challenges not only firms and other private sector organisations but also localities interested in sustaining economic prosperity. Understanding the drivers of the distribution of economic activity in space can therefore be key to sustainable regional development, especially when it comes to the issue of local industrial centres, i.e. locations of heightened economic activity.

In the context of modern-day industrial centres (e.g. Silicon Valley), theoretic analysis is based on those empirically derived observations about apparent benefits from co-location that were first voiced by Alfred Marshall (e.g. in the 8th Edition of 'Principles of Economics', 1920). Existing and modern Marshallian agglomeration theory is focussed on an elaboration and further development of the mechanisms leading to benefits from co-location. As a result, this body of theory investigates the success factors of an existing agglomeration (or 'cluster') of firms at a specific point in time.

Within this context, Kerstin Press takes up an issue that is neglected in much of the theoretic literature on clusters while being very prominent in the real world: Do the dynamics of agglomeration, change and adaptation follow a life-cycle from emergence to decline, or is the development of existing clusters over time a non-deterministic one? Her study addresses this aspect by analysing the adjustment of existing clusters to exogenous shocks in their greater economic environment. It proceeds by investigating the success factors for regeneration and survival of the cluster, focussing on the effect of the cluster's architecture on the self-organisation processes taking place among its organisations when they adjust to exogenous shocks.

The analysis of such dynamic processes with methods from mainstream economics is problematic due to the dominance of general equilibrium models in the field. Since the focus of the present study is not on comparing initial and equilibrium states but on the development process following an external shock, Kerstin Press adopts a new methodological approach. She develops a theoretic model of clusters, change and adaptation that is based on the N/K model of complex systems (Kauffman 1993). Taking the existing insight on the dynamics of N/K systems, propositions on the influence of cluster architecture on adaptability are derived. In a second step, these propositions are tested by simulations comparing the adaptability of clusters with given differences in their architecture regarding the degree of division of labour on the one and the governance structure on the other hand. Both aspects (division of labour, governance) were found to matter for cluster adaptability in previous empirical studies but general causalities for their role in cluster adjustment could not be derived from individual case studies. Within the all else held equal perspective adopted in the simulation model, one explanation for how and when division of labour and governance matter for cluster adjustment is found. While the model faces several limitations, these are more to be seen as possible areas for future research.

The present study opens up new avenues in cluster research. Rather than just focussing on emergence or existence, clusters are viewed as entities underlying a dynamic development. The analysis of their adaptability to external changes constitutes a first important step towards a non-deterministic perspective on cluster development. The main contribution of Kerstin Press lies with a successful bridging of the gap between empirical observations and theoretical insights regarding clusters and adjustment. By providing a model able to explain stylised empirical findings while being based in the body of theoretical knowledge on clusters, important causalities for future theoretic and empirical research have been derived. The sound basis in the existing theoretic knowledge is also evident in an excellent presentation of the state-of-the-art literature on agglomeration and cluster research found in the present study.

Kerstin Press' dissertation offers a promising methodological approach, which – hopefully – encourages others in the field to transfer and apply the methodological tools available within and outside mainstream economics to other dynamic economic phenomena.

Duisburg, January 20th 2006   Günter Heiduk

Acknowledgements

This project would not have been possible without the generous advice and support I was fortunate enough to receive from a number of people and institutions.

I began working on my PhD dissertation in October 2001 at the Chair for International Economics, University of Duisburg-Essen. From the very start onwards, my thesis advisor Prof. Dr. Günter Heiduk took a very active interest in the project. He was virtually always available for discussions about ideas and (often unfinished) lines of argument and gave me considerable degrees of freedom for linking up with the international scientific community. I have also come to value his swift reviews of thesis chapters as well as the final monograph over the years. In sum, being his PhD student has been an enjoyable experience.

I would also like to thank Prof. Dr. Markus Taube for being the second referee of my dissertation. Despite their dense schedules, my examining committee comprising Prof. Dr. Günter Heiduk, Prof. Dr. Markus Taube as well as Prof. Dr. Dieter Cassel and Prof. Dr. Torsten Gerpott made time available for me to defend the thesis on the final day of my PhD studies. I greatly appreciate your help in meeting this very date.

During my PhD studies, I was employed as a research assistant at the Chair for International Economics, University of Duisburg-Essen. I would like to thank my colleagues for one great work environment. Dipl.-Oec. Christian Schabbel is one of the most fun people to work, travel and share an office with. Our secretaries (in sequence) Ms Brigitte Dunkel, Ms Marianne Appelt and Ms Nicole Jaschinski managed many daily details I am still not fully aware of and kept everyone supplied with the bare necessities – be they office material or fresh coffee. Several 'generations' of student assistants including André Böhmert, Andreas Eickel, Sonja Fischer, Jörg Knips, Nadja Kremser and Stefanie Lenz conquered internet resources as well as the local library and inter-library loan systems to assemble my bibliography (among other things).

On my ventures outside the University of Duisburg-Essen, I met a number of people that critically affected my research agenda. The project benefited immensely from discussions with colleagues at different conferences and workshops. Among the many venues attended that shaped the thesis in often intricate ways (when seen in retrospect) some were key in finalising the present study, including: The Research on Organizations, Coordination & Knowledge (ROCK) Workshop '*NK Model and Applications to Economics and Management*' (Trento, Italy), the Danish Research Unit on Industrial Dynamics (DRUID) PhD Winter Conference '*Industrial Evolution and Dynamics*' (Skørping, Denmark), the Consortium for Competitiveness and Collaboration '*The CCC's 12th Annual Colloquium for Doc-*

toral Student Research' (Berkeley, CA) and the DRUID 10th Anniversary Summer Conference '*Dynamics of Industry and Innovation: Organizations, Networks and Systems*' (Copenhagen, Denmark).

In 2004 and 2005, I had the chance to attend the European Doctoral School in the Economics of Technological and Institutional Change (EDS – ETIC). In particular, I would like to thank the local organisers at University Louis Pasteur (Strasbourg, France), Patrick Llerena and Monique Flasaquier as well as the teaching staff and fellow PhD students involved with the ETIC 2004 Spring Session and the Simulation in Evolutionary Economics (SIME) Eurolab Course in 2005. Both in- and outside class discussions have been incredibly rewarding.

Regarding the PhD thesis itself, several individuals have made outstanding contributions (in order of appearance). Philip Cooke was an invaluable source of ideas on biotechnology-related research and a very effective editor to work with. Thomas Brenner introduced me to the idea of life cycles in clusters and thereby indirectly contributed to a more theoretic orientation of my research project. In a later stage of the PhD, I met Peter Maskell who became an inofficial 'second' advisor to my thesis. This started with discussions at a couple of conferences and culminated in a stay as visiting researcher at the Institute for Industrial Dynamics and Strategy (IVS), Copenhagen Business School. The generous funding provided by the European Union's Marie Curie Programme is also gratefully acknowledged. It is probably fair to say that the thesis would not look the same without the time spent at IVS. When working out the details of the theoretic model, I strongly benefited from discussions with Thomas Brenner, Koen Frenken, Peter Maskell, Olav Sorenson, Marco Valente as well as my colleagues at IVS. Not being much of a programmer myself, I am particularly indebted to Marco Valente for his assistance and guidance in turning theoretic reasoning into a workable simulation model. Financial support for publishing the thesis by the DVF (Duisburger Volkswirtschaftliches Forschungsseminar) is also gratefully acknowledged.

On a final note, I would like to pay tribute to the indirect supporters of the present study. Very early in my PhD, I used to think that acknowledgements to friends and family were just a lovely tradition adhered to by most authors. My ignorance on the significance of the contribution of one's social network to a long-term project like the PhD was very soon at an end. I would like to start by thanking my parents Barbara and Hans-Winfried Wolter as well as my brother Niels who set the foundations for a path I am only beginning to walk. My grandfather, Karl Hoffmann, cheered me up with more anecdotes on academic life than I can recount here. Last but certainly not least, I would like to thank my husband, Achim Press, for enduring more than one thesis-induced mood swing. I could always rely on your faith when mine failed.

I therefore dedicate this monograph to my extended family.

Duisburg, January 20th 2006   Kerstin Press

Table of contents

Part I The rationale for studying cluster dynamics 1

1 Introduction: Clusters, change and adaptation 3
    1.1 A Janus-faced phenomenon: Cluster prosperity and decline 4
        1.1.1 Definition and unit of analysis 7
        1.1.2 Cluster dynamics: On emergence, endurance and exhaustion.............9
        1.1.3 Decline or adaptation: Agent activities and cluster architecture 12
    1.2 Goals and contribution of the study 13
    1.3 Course of the study 16

Part II Literature review – The benefits of co-location 19

2 Stability and change: Driving cluster development.......................21
    2.1 Novelty, uncertainty and transaction costs: Industry-location life cycles..23
    2.2 Symbiosis, habitat and external events: Dynamics of firm populations 25
    2.3 Externalities and trade costs: The New Economic Geography 27
    2.4 Increasing returns and firm location choice: Spatial path dependence 31
    2.5 Summary and critique.........................35

3 The nature of the beast – On the notion of agglomeration externalities......37
    3.1 The advantages from co-location.............................39
        3.1.1 It all started with Marshall: Agglomeration economies.....................41
        3.1.2 Beyond Marshall: New developments in the cluster literature..........43
    3.2 On Districts, Porterian 'Clusters', Innovation Systems and Milieux 47
        3.2.1 Italian Industrial Districts 47
        3.2.2 (Porterian) Clusters...50
        3.2.3 Regional Innovation Systems 51
        3.2.4 Innovative Milieux 53
    3.3 Summary and critique...54

Part III Towards a complexity perspective on clusters 57

4 Clusters, change and adaptation: Sticky places in slippery space?............. 59
4.1 The nature of clusters: Agents, interdependence and co-ordination 60
4.2 Adaptation: Decentralised problem solving by interdependent agents 64
4.2.1 Individual activities or cluster properties: Drivers of adaptation 65
4.2.2 Reality bites: Success factors in adaptation.................................... 68
4.2.3 Agents, interdependence, adjustment: Cluster self-organisation........ 73
4.3 The nature of cluster adaptation and resulting model requirements 75

5 Modelling adaptation in clusters – The promise of complexity theory........ 77
5.1 Clusters as complex adaptive systems 78
5.1.1 Systems and complexity theory: An overview 78
5.1.2 Are clusters complex adaptive systems? The issue of agency 82
5.2 Agents, interdependence and fitness: Introducing the N/K model............. 84
5.2.1 The importance of directed interdependencies................................. 85
5.2.2 Measuring success by fitness landscapes 86
5.2.3 Bifurcations or perturbations in the fitness landscape...................... 90
5.2.4 Structure, search, selection: Division of labour, co-ordination 91
5.3 Clusters as N/K systems: Parameter definition and system dynamics....... 92

Part IV Model development – Clusters as complex adaptive
     systems... 93

6 Micromotives and macrobehaviour – Dynamics of N/K systems................. 95
6.1 Agent dynamics: Structure, Search and Selection 97
6.1.1 Structure: The nature of fitness landscapes..................................... 97
6.1.2 Search: Landscape exploration by mutation.................................... 100
6.1.3 Selection: Evaluation of modification fitness.................................. 102
6.2 The dynamics of agent groups 104
6.3 Co-evolving agent populations: Self organisation and performance 105
6.4 The behaviour of N/K systems: Agent, group and aggregate dynamics .. 108

7 Clusters as co-evolving N/K systems.. 109
7.1 Parameters, agent, group and system dynamics.. 109
7.1.1 Model parameters: N, K, C and S 111
7.1.2 Model setup: Agents, elements and interdependence........................ 113
7.1.3 Search, test, selection: Agent dynamics 116
7.1.4 Bidding for representation: From agent to group behaviour 118
7.1.5 Deforming landscape subsets: The dynamics of co-evolution 118
7.2 Accommodating division of labour and mode of co-ordination 119
7.2.1 Fitness landscape structure: The role of division of labour............. 120
7.2.2 Strategy selection: The role of inter-agent co-ordination 125
7.3 Number, optimality and spread of modifications: Driving adaptability .. 130

Part V Division of labour, co-ordination and cluster adaptation 133

8 Clusters, change and adaptation – Simulation results 135
  8.1 Fitness landscape reconfigurations (bifurcation) 136
  8.2 Fitness landscape shift (perturbation) 147
  8.3 The tragedy of collectivists 153
  8.4 Summary 159

9 Model contribution, limitation and avenues for future research 161
  9.1 Implications of findings 162
    9.1.1 Division of labour and co-ordination: The causalities.....................162
    9.1.2 Explaining empirical trends: Districts and Silicon Valley...............164
  9.2 Model contribution 166
  9.3 Model limitations 168
  9.4 Avenues for future research 171
  9.5 Conclusion 172

Appendices 175
  A1 Symbiosis, habitat and change 175
  A2 Clusters as core-periphery structures 176
  A3 Location choice, path dependence and clusters 178
  A4 Model parameter values and fitness landscapes...........................182
  A5 The simulation model 189
  A6 Cluster adaptation to change – Results overview...........................203

List of figures 223

List of tables..225

References 227

Part I The rationale for studying cluster dynamics

A district which is dependent chiefly on one industry is liable to extreme depression, in case of a falling-off in the demand for its produce, or of a failure in the supply of the raw material which it uses.

Alfred Marshall 1920, p. 273

Who will win the globalisation race? For most nations, this question is very closely connected to the success of its firms and industries. Insights into the determinants of firm competitiveness have thus always been of chief interest – not only to policy makers, but also to economic scientists. In this context, the concept of 'Clusters' has attracted considerable attention since the early 1990s. This is attributable to the fact that clusters not only enhance firm competitiveness but also increase their spatial embeddedness. The concept argues that spatial concentrations of companies in the same or related industries are beneficial to firms. Due to their co-location, companies are able to forge trust based relationships, not only with other firms but with other important regional players (such as government institutions, local buyers, local universities and so on). Therefore, firm innovations spread faster, products become more specialised and are also upgraded at a quicker pace within than outside such clusters. As these positive cluster effects are based on local linkages between actors, the concept also offers an explanation of the phenomenon that despite increasingly global markets, companies in one industry tend to locate together (the 'globalisation-localisation paradox').

What is often forgotten in the euphoria about creating and enhancing regional prosperity through clusters is that areas facing severe structural problems today were thriving clusters in their time. However, as the technological evolution progressed, these regional structures proved unable to adapt and were rendered obsolete. At the background of an increasing number of countries pursuing some sort of strategy to create the next Silicon Valley, the question of whether clusters exhibit a life-cycle – or whether their deterioration can be avoided in a world of increasingly fast technological developments has to be asked. This study addresses the aforementioned issue by investigating *whether, how and when agents in clusters can adapt to adverse change events in their environment and what factors at the cluster level may assist them in doing so.* By comparing adaptive performance of clusters exhibiting different architectures, the model results shed light on empirical findings by linking them with theoretic research. Moreover, the findings indicate that the cluster's architecture that has evolved throughout its past history can be better or worse for its future survival. Finally, the study constitutes a first step towards a more dynamic perspective in cluster theory.

1 Introduction: Clusters, change and adaptation

This study deals with the factors shaping the outcome of adverse change events in industrial clusters. Such periods of crisis are likely to occur throughout the development of virtually any cluster when events affecting their area or industry call for a change in current practice to retain competitiveness. In many instances, a failure to adapt to such events has meant economic decline for the local industry with often severe repercussions for the prosperity of entire areas as is witnessed by the many old industrial regions throughout the world. However, other cases exist in which agents in the cluster were able to accommodate the change event, thereby allowing the local industry to survive. By dealing with the factors underlying success or failure of cluster agent adaptation, this study thus contributes to our understanding of the circumstances under which a decline of the local industry can be avoided. The relevance of this research question, i.e. *whether, how and when cluster agents can successfully adapt to change events* emerges from three aspects. First, there is significant empirical evidence of and political as well as theoretic attention dedicated to the phenomenon of clusters. Second, with the accelerating dynamics in technological and industrial developments characterising the modern economy, an increase in the dynamics of the spatial distribution of industries can be expected over time, thereby giving rise to new and challenging existing clusters at a faster pace. Third, empirical evidence on existing clusters' adaptation to change has shown some striking resemblances between surviving clusters, which merit further investigation.

This chapter proceeds as follows. The following section sets the stage for the subsequent analysis. It outlines the background to the study pursued here by showing a rationale for studying cluster dynamics based on the prominence of the phenomenon as well as historic insight about the possibility and consequence of cluster decline. Sect. 1.1.1 then provides the definition of clusters employed, distinguishing it from related terms like agglomeration or districts and outlining its difference from administrative spatial constructs like regions (areas) or cities. Sect. 1.1.2 investigates the role of change events for cluster dynamics in more detail. It finds that the stability of clusters is attributable to the effects of agglomeration externalities supported by a local culture while emergence and decline relate to (often external) change events. However, adverse external events do not imply an inexorable road to cluster decline. As is found in Sect. 1.1.3, adaptation to and survival of change is possible. Reviewing empirical evidence of survivor cases then shows a link between two architectural factors in clusters (division of labour, co-ordination) and adaptability. It is this link that will be investigated in more detail through a theoretic model and simulations. Several contributions to the existing literature emerge from this (Sect. 1.2). First, model results provide a more conclusive analysis of the role of both factors for cluster adaptability which can explain some of the empirical observations. Second, the study shows that the Marshallian perspective on clusters can be aligned with dynamics. Finally, the identification of adaptive cluster architectures can act as guidance for future empirical and theoretic research although the focus of the model implies that there will be *no one-to-one relationship* between cluster architecture and survival. Sect. 1.3 outlines the course of the study pursued here.

1.1 A Janus-faced phenomenon: Cluster prosperity and decline

The clustering of industries understood as a non-random spatial concentration of economic activity (Ellison and Glaeser 1997) is a phenomenon that has received increasing empirical, political and theoretical attention. Based on ideas first advanced in Alfred Marshall's 'Principles of Economics' (1890), the existence of clusters is justified by advantages from co-location due to agglomeration externalities. The latter are supported by a local culture, i.e. a set of formal and informal rules of the game established in the area that co-ordinate agent activities as to produce better results than would be available to isolated firms.

Empirically, the uneven distribution of resources, production factors and especially economic activity over space is and has been a defining feature of economic reality worldwide, even despite drastic falls in transportation and communication cost (Puga 1998; Puga and Venables 1996; Quah 2002). Many studies investigating the phenomenon have yielded evidence on the propensity of actors in very different industries and nations to concentrate spatially. Although this literature has been criticised as being mainly based on 'stories' rather than statistical investigation at a more macro level (Head et al. 1995, p. 223), work in the quantitative vein also identifies a significant trend towards clustering at the level of nation states.[1]

The real world evidence of the existence of successful clusters and their benefits to the host region and nation has led to an increased policy interest that continues to spread within the OECD world and beyond (Malmberg and Maskell 2002, p. 431): Policy actors in many nations have started to embrace the view that creating clusters in different industries is the main option of regional policy in order to gain or sustain regional competitiveness and thus survive the 'globalisation game' (Lagendijk and Cornford 2000; Lundvall 2002; Maillat 1998b).[2]

---

[1] For case studies see Brusco 1982, 1986, Brusco and Righi 1989; Dei Ottati 1994c; Goodman et al. 1989; Lazerson 1990, Saxenian 1991 and Staber 1996a. Statistical investigation are found in Ellison and Glaeser 1994, 1997, 1999; Krugman 1991a; Hanson 1996a; Maggioni 2002 or Malmberg and Maskell 1997.

[2] The role of successful clusters for their host area is emphasised by Bresnahan and Malerba 1999; Hirschmann 1967; Krugman and Venables 1995; Myrdal 1957; Perroux 1950; Porter 2000; Rosa and Scott 1999; Scott 1988, 1998; Sforzi 1990 or Storper 1997.

Alongside empirical and political interest in the cluster phenomenon, Marshall's ideas on its underlying mechanisms have received renewed and growing theoretic attention within mainstream economics since the early 1990s. This *rediscovery of space* was propelled by Porter's book "The competitive advantage of nations" (Porter 1990) as well as Krugman's (1991b) paper marking the introduction of a strand of research now known as the "New Economic Geography". The increasing theoretic interest of economics in *"where economic activity takes place – and why"* can be attributed to its importance for other core aspects in the field, such as *"urban economics, location theory, or international trade"* (Fujita et al. 1999, p. 12).[3]

An aspect that is often forgotten in the political and theoretic euphoria about the benefits of successful clusters is the fact that the very uneven distribution of industry activity that can lead to clusters is anything but unchangeable. In economic history, industries have been shown to migrate between regions and countries (see Courtney 1878; Engländer 1924; Maunier 1908, 1909; Quelle 1926; Ritschl 1927; Schlier 1922; Schumacher 1910 or von Beckerath 1918-1922), a trend confirmed by more recent evidence (e.g. Audretsch and Cooke 2001; Bluestone and Harrison 1982; Bresnahan et al. 2001; Carney 1980; Courlet and Soulage 1995; Dumais et al. 2002, Gertler 1996; Mathias 1983; Massey and Meegan 1982 or Rosenberg 2002). However, there is no general pattern in industry dynamics: Some sectors disperse spatially while others stay clustered. What emerges is, however, that developments at the local or the industry level can lead to a shift in the latter's distribution over space. This shift can be introduced by a number of factors including new technologies or changes in demand as well as exhaustion of natural resources. With the modern economy's increased pace of development, especially the first two kinds of change are likely to become more frequent, thereby increasing the number of challenges any existing cluster has to face.

At the level of the region, this potential for a change in industry distribution means that former agricultural areas like the Silicon Valley can come to host the world-leading cluster of firms in the computing industry. At the same time, *"what once was a leading centre of dynamism within a given line of business [can end] up as an 'old industrial region', facing great problems of renewal and finding itself out-competed by firms located elsewhere"* (Malmberg and Maskell 2002, p. 432) as is witnessed by the many areas facing severe structural problems today (such as the old industrial centres of the Ruhr in Germany or Detroit in the US). The consequences of this cluster decline on their host regions are often harsh: *"It is distressing [...] to witness the havoc which has been made in some picturesque valleys in Lancashire by the pressure of modern emulation and competition. Factories and cottages closed and untenanted, many of them unroofed and in ruins,*

---

[3] The fact that space was rediscovered by economists is not to deny the long-standing tradition of the topic within the field of economic geography. This study will not address the ongoing dispute between both disciplines as to how the spatial distribution of economic activity should be studied. See the controversy between Krugman 1991b and Martin 1999 or Martin and Sunley 1996. Scott 2004 provides a more balanced account of the contributions of geography and economics to the study of clusters.

meet the visitor at almost every turn, and give some indication of the great sacrifice of capital which must have been made before the present hopeless condition was reached. A tornado can hardly vie with the forces of competition in power to level and destroy" (Ross 1896, p. 267). The benefits of clusters to their host area are thus under a constant threat of being lost because clusters can be extinct by events that change the spatial distribution of their industry.

Empirical evidence on the development of clusters has revealed that despite the dynamics in the spatial distribution of industries, there is no such thing as a deterministic life-cycle running from emergence to exhaustion. Depending on the response of local agents, change events may not shift the industry away from an existing cluster. Instead, local decline can be avoided if agents in the cluster are able to accommodate the new situation. This adaptation proceeds with different time-horizons and mechanisms. In a first step, following the change event, the immediate response by local agents in an existing cluster involves a bid to adjust to the new situation by changes in their strategies. This *adaptation by self-organisation* will be shown to exhibit a number of interesting peculiarities thanks to the nature of clusters as they are understood in this study. Depending on the success of local agents in this first step, a second phase of adjustment can then involve *adaptation through arbitrage*. The foundations underlying the outcome of adaptation in this phase are laid in the first one:

- If some local actors survive the first phase, arbitrage can involve a consolidation of the local industry through emergence of new organisations and exit of unsuccessful ones. In a more long-term perspective, a surviving local industry can also come to form the basis of a new sector by acting as a source of technology or demand for new industry firms (Lee et al. 2000; Henton et al. 2002 or Sturgeon 2000). This avenue towards *industrial substitution* (Gabe 2003; Lane 2002; Swann 1996) depends on survival of the old sector.
- If local actors cannot self-organise a response to the change event that is successful enough for the survival of parts of the cluster, adaptation can only proceed at a higher level of aggregation. Investigating the cluster's host area, arbitrage could involve a compensation for losses in the local industry's activity by increases in other sectors. In a long-term perspective, structural change by transition towards new industries in the region could also be enabled by non-industrial incubators such as research institutions.

The survival of a specific clustered industry is therefore conditional on how well its agents can adapt to change, first by self-organisation and second through arbitrage. With respect to the cluster's host area, the success or failure of adaptation in a local industry can then have very immediate and more long-term repercussions by influencing economic prosperity as well as the possibilities for structural change in a region through industrial substitution (Venables 2001). In a way, the outcomes of the first phase of adaptation, when an existing cluster is challenged by a change event, therefore have immediate and long-term repercussions, both for the cluster itself as well as its host area.

The remainder of this chapter sets out the foundations upon which the study builds. It starts by outlining the unit of analysis, defining clusters as spatial concentrations of economic activity characterised by agglomeration externalities. Sect. 1.1.2 then investigates the role of change for the development of clusters. Sect. 1.1.3 looks into the meaning of adaptation within the context proposed here. It asserts that the nature of clusters matters for the outcome of adaptation to change by exerting architectural constraints on agent activities: Thanks to the existence of positive and negative externalities, specific agent strategies work better or worse in the presence of others' activities than they would in isolation. In addition, many clusters develop a set of rules on acceptable business practice (the local culture) that might restrict the available avenues for agent adaptation. Two aspects emerge from this:

- Agent adaptation does not equal cluster adaptation.
- Some architectural properties in clusters may be more suitable in steering individual adaptation towards good collective outcomes.

Sect. 1.2 derives the goal of the study conducted here as well as its contribution to the existing literature. The final section then illustrates how the remainder of the study will tackle the question of whether, when and why two stylised empirical (cluster-level) factors regarding the extent of division of labour and the mode of co-ordination can matter for cluster adaptation by gearing individual actions towards suitable outcomes for the entire cluster.

1.1.1 Definition and unit of analysis

A first issue to be addressed concerns the definition of the term *cluster* to be adopted here as well as its delimitation from other spatial constructs such as agglomeration or administrative units like cities and regions (areas). In meteorology or the natural sciences, agglomeration is the result of a process bringing smaller particles together to form a larger mass. In the context of economic geography, the term refers to the spatial concentration of people or economic activity (Malmberg and Maskell 2002, p. 430). Agglomeration can therefore mean the concentration of human settlements in cities or town areas or – as it will be investigated here – the concentration of economic activity in space. *Agglomerations* in the most general sense will therefore be defined as the locus of heightened economic activity.

The necessary condition for the emergence of agglomerations is thus an uneven spatial distribution of economic activity. The latter can take on two different forms depending on the 'boundary condition' chosen. On the one hand, when delimiting agglomerations according to administrative or political 'frontiers', very diverse industrial structures can be observed at different spatial levels including cities, administrative regions or entire nations. On the other hand, a somewhat different picture emerges when the sector is chosen as the boundary condition. Again, more specialised concentrations of firms in one or few related sectors could be observed at different spatial levels. At the global level, a few countries can sometimes be found to host most or all leading firms in an industry (Porter 1990) and within individual countries, industries can be found to agglomerate in smaller areas. It is the latter case of *industrial agglomerations*, as concentration of firms in one or several related sectors within a confined geographic space (Porter 1998) that will be investigated in more detail here. Such agglomerations can but need not coincide with the administrative boundaries defined by cities or regions. Rather than by administrative limits, they are defined by the relationship between their constituent firms.

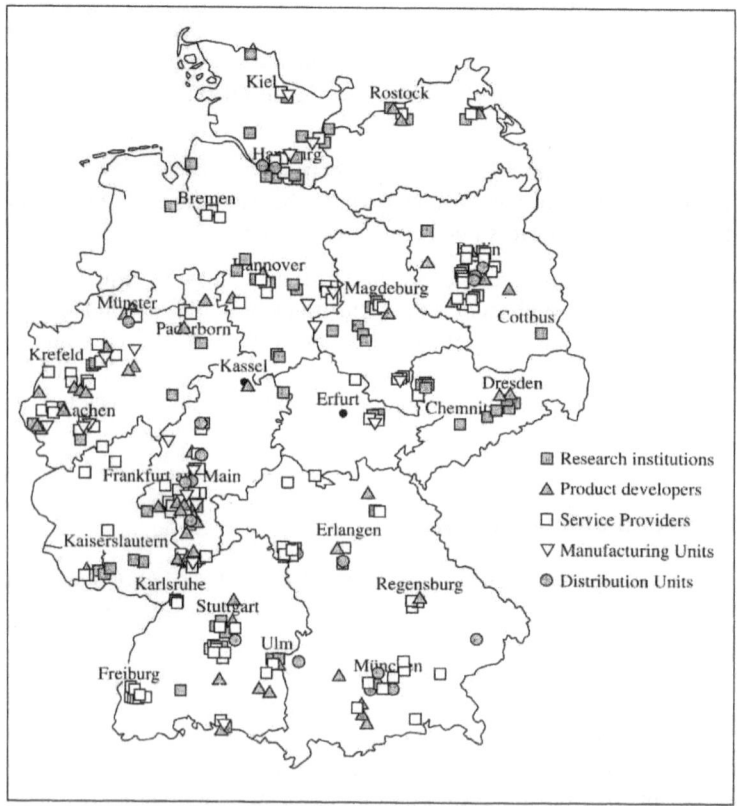

Fig. 1.1. Spatial industry distributions - German pharmaceutical biotechnology
Source: Wolter 2005, p. 132 (with permission from Edward Elgar Publishing Ltd.)

The propensity of many different industries to concentrate spatially has been witnessed empirically (for one example see Fig. 1.1.). Any uneven distribution of industry activity is in turn a direct consequence of the location choice of firms: An industrial agglomeration can only emerge if (a sufficient number of) firms choose the same area as their site of operation. The trend towards a co-location of firms in an industry could itself be attributable to many different causes including location intrinsic factors (e.g. climate, raw materials) or industry characteristics leading to proximity benefits in the most general sense. If an industrial agglomeration was no more than the result of identical firm location choices, its emergence and development could be described as a consequence of the former. Furthermore, the 'success' of the area would be equal to the 'sum' of the competitiveness of its firms.

However, firms not only choose a locale as the site of their operations but also perform different kinds of business activity there. This in turn can lead to additional benefits and costs to being located in the area, which are labelled *agglomeration externalities* (see also Chaps. 2 and 3). Their existence enables the emergence of a local concentration of industry activity exceeding the level warranted by location intrinsic or transport cost related factors (see Ellison and Glaeser 1994; 1997 and 1999; Glaeser et al. 2002 as well as Roos 2002a). At the same time, negative agglomeration externalities constrain the 'carrying capacity' or the total number of firms an area can host. These non-random agglomerations characterised by agglomeration externalities will be termed *clusters* for the remainder of this study.

The definition of clusters adopted here is intentionally broad and includes any spatial concentration of agents involved in a specific set of economic activities that exhibits agglomeration externalities. While agglomeration is thus a necessary condition for the existence of clusters, it is not a sufficient one. Agents in the cluster furthermore have to incur benefits and costs to being there that are a consequence of the activities of other actors (agglomeration externalities). The existence of such externalities implies interdependence between the activities of local agents that will constitute a key factor in their adaptation to change events (see also Chap. 4). Moreover, the definition employed here is general enough to allow for different extents and types of agglomeration externalities. As a result, the constructs proposed by post-Marshallian concepts (Italian Industrial Districts, Porterian Clusters and Regional Innovation Systems as well as Innovative Milieux; see 3.2) constitute specific variants of clusters since they emphasise the roles of different agglomeration externalities for different firm activities. With respect to higher levels of aggregation, clusters are components shaping parts of the industrial structure of higher-level administrative spatial units like cities or regions.

1.1.2 Cluster dynamics: On emergence, endurance and exhaustion

The forces determining the spatial distribution of economic activity over time can have important repercussions on the development of clusters as well as their host regions or nations. The question of cluster dynamics can thus be addressed from two different angles. The first involves a focus on their changing role in regional and national economic development (Myrdal 1957; Perroux 1950; Puga 1998), although the promise of clusters for prosperity is often exaggerated (Amin and Robins 1990 versus Porter 2000). The second perspective lies with investigating the factors shaping the development of the cluster itself, i.e. the drivers of emergence, endurance and exhaustion. Adopting this perspective on many existing clusters shows that their development proceeds through phases of continuity and change (Fig. 1.2.).

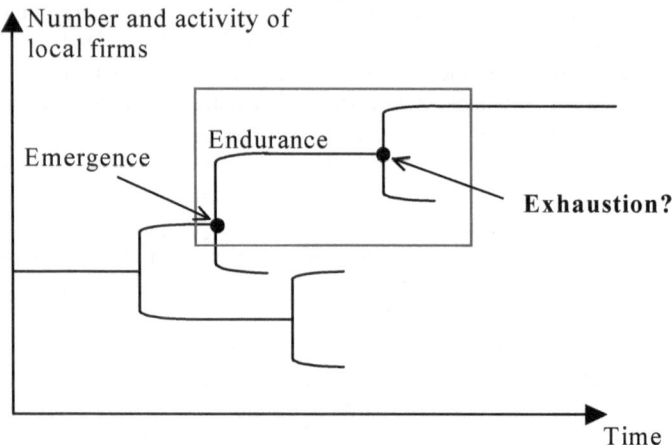

Fig. 1.2. Continuity and change in cluster development
Source: Adapted from Allen 1997, p. 18 (with permission from Gordon and Breach Science Publishers)

Cluster emergence is often triggered by events that make a natural or social asset of the area an important location factor for an industry or that let to a (group of) local agent(s) engage successfully in a specific sector (e.g. Feldman and Schreuder 1996). The subsequent growth of the cluster proceeds through start-up and spinout activity inspired by the success of pioneering firms, an increasing division of labour, a consolidation of the industry (mergers, acquisitions or firm exits) as well as the attraction of resources from outside the area (people, capital or entire organisations). Throughout this phase of cluster endurance, agglomeration externalities exert a stabilising influence: On the one hand, agglomeration economies increase the competitiveness of clustered firms compared to that of isolated organisations. On the other hand, negative externalities limit the growth of the cluster. Otherwise, any industry would end up in the same location: "*[...] one might wonder why the rising centre does not go on killing out its small rivals over larger and larger areas and appropriating their business. But trees do not grow up into the sky, nor does an industrial centre expand till it absorbs the custom of the globe. For this, there are two chief reasons. One is that, as a place becomes more of a centre, its special advantage tends to disappear. [...] The other reason is that, as industry concentrates, the radius of the territory from which its materials and the subsistence of its dependent population are drawn, and of the territory over which the finished product is distributed, increases; the average cost of transportation per unit of industry grows, until its growth neutralizes the economies of further concentration*" (Ross 1896, pp. 263-264). The trade-off between positive and negative agglomeration externalities therefore determines the maximum size of the cluster, i.e. the area's *carrying capacity* (Chap. 2).

While they can create a cluster, change events are also responsible for its exhaustion. Again, history shows that most clusters run into problems at some point in their development when events affecting the area or the local industry call for a restructuring of current practice (Malmberg and Maskell 2002, pp. 431-432). These developments can turn the one-time advantages of being in a location into liabilities. This aspect of extinction is a lot less prominent in the more recent literature than the previous two. Whenever extinction arguments have been put forth, they are attributed to local (internal) or industry-wide (external) developments. The former, i.e. internal sources of exhaustion include developments specific to the cluster, which reduce agglomeration economies or increase congestion cost thus making location in an area less advantageous. They could lie with a loss of rivalry, the emergence of internal or regulatory inflexibilities (Porter 1990) as well as increasing opportunistic behaviour (Maillat 1998b). At the same time, external developments affecting the industry or its host area are a second source of challenges to existing clusters. Frequently stated examples include technological discontinuities, changes in quality and quantity of demand, changes in factor proportions of production, improved or cheaper transportation[4] as well as modifications in host nation industry policy or legislation (Porter 1990; Ritschl 1927).

The focus in the study conducted here will lie with cluster agent's adaptation to these external challenges, i.e. changes that did not originate with the activities of local agents but that impact upon them like an outside shock. Two arguments speak for this perspective. First, internal developments like the ones outlined before already reduce the benefits from co-location in a cluster. As a result, it is far more vulnerable to any external challenge paralleling its internal development than a well functioning one. By studying the adaptation of agents in functioning clusters to external change events, one therefore arrives at a 'best case' scenario. The implications of the latter will hold even more strongly in worse cases. Second, the increasing pace of development with respect to technological possibilities or demand preferences implies that this type of change event is likely to become more important in the future, increasing the number and frequency of external challenges that any existing cluster will have to face.

With the advent of an external change event challenging the cluster, the issue of decline or adaptation emerges. With respect to their scope, some of these external events (new technologies) correspond to industry-wide bifurcations in the Schumpeterian sense of *creative destruction* (Schumpeter 1934b) whereas others (changes in demand quantity) correspond more closely to smaller perturbation events. Both types of external shock can however result in a *local bifurcation*, i.e. in a situation in the cluster's development where its current trajectory is disrupted and multiple future avenues of development emerge. At the level of the cluster, a local bifurcation can have very different outcomes: Adaptation of existing actors or depopulation and economic stagnation (Tichy 2001). The immediate response of cluster agents to the local bifurcation (see Sect. 1.1) can therefore influence the future development of the cluster or more precisely whether the result of the local

---

[4] In traditional location choice theory, the exhaustion of an industry site is brought about by developments affecting the role and attractiveness of different location choice factors. A major role in this context is attributed to transport costs (see Engländer 1924; Ritschl 1927; Sax 1918-1922 and of course Weber 1909).

bifurcation is decline or (some degree of) successful renewal. The question this study thus seeks to address is *whether and when agents in an existing cluster are more likely to adapt successfully and thereby survive a local bifurcation.*

1.1.3 Decline or adaptation: Agent activities and cluster architecture

As will be seen in the following chapter, those theories dealing with cluster dynamics hypothesise that any change in the parameters underlying the spatial distribution of an industry produces a deterministic reaction towards increased agglomeration or dispersion. The second aspect (dispersion) implies decline for any existing cluster. Empirically, however, local bifurcations have not elicited the same response. In some instances, they have meant decline (see Grabher 1993; Isaksen 2003; Massey and Meegan 1982; Meyer-Stamer 1998; Müller et al. 2005) while in other cases, successful adaptation has been observed (e.g. Bellandi 1996; Bresnahan et al. 2001; Bresnahan and Malerba 1999; Cooke 1997; Dei Ottati 1996; Glaeser 2003; Guerrieri et al. 2001; Lee et al. 2000; Paniccia 2002; Saxenian 1994; Schamp 2000; Schmitz 1999 or Tiberi Vipraio 1996). This evidence highlights that adaptation is possible – yet not under all circumstances.

The question of cluster survival then depends to a considerable extent on the first phase of adaptation, i.e. the self-organisation processes between local agents that directly follow a change event. The outcome of this phase sets the basis for any subsequent adaptation by arbitrage involving the emergence of new agents as well as changes in the area's industrial mix. While this is not to say that failure of existing local actors to adapt in the first phase means that all is lost, reversing an adverse development trajectory created by their reply becomes more difficult, the longer this trajectory exists (Prigogine and Stengers 1984; Prigogine 1997; Sect. 4.2.3). At the same time, success in the first phase of adaptation does not guarantee cluster survival as multiple influences in the second stage matter as well (for detail see the following section).

When investigating the factors impacting on the success of agent's self-organisation after a local bifurcation, the nature of clusters allows for two different perspectives. The first would regard the drivers of the quality of agents' adaptation, i.e. when individual organisations are more or less responsive to change depending on their degree of inertia. As will be elaborated in more detail in Sect. 4.2, the success of agent activities in clusters is however also influenced by the architectural features of the cluster itself. On the one hand, agent adjustment depends on the activities of others due to positive and negative agglomeration externalities. Moreover, in the presence of a local culture acting as a co-ordination mechanism between cluster agents, the latter are restrained in their behaviour insofar as defection from the established (and short-term unchangeable) local rules of the game can induce punishment by other actors. An investigation into individual agent adaptation would therefore not tell the full story of adaptation from the cluster's perspective.

A second view on the success factors in cluster adaptation therefore lies with the role of specific cluster-level (architectural) properties in steering individual agent activities to good collective outcomes. Empirical research on the nature of surviving clusters has found a number of interesting insights with respect to what these cluster-level success factors for adaptation may be. A comparison between the responses of the computing industry in the Silicon Valley and Boston's Route 128 following the introduction of the microcomputer (Saxenian 1994) emphasises the flexibility advantages of small, networked producers over large integrated ones. The extent of division of labour thereby seems to matter for the outcome of cluster agent adaptation processes. A second strand of research investigating surviving Italian districts (Cainelli and Zoboli 2004a; Lombardi 2003 or Maggioni 2004) finds evidence of an overreaching development away from egalitarian networks towards more hierarchical structures. The intuition behind this can be found with Sebastiano Brusco's idea on the difficulties of accommodating change in districts with many equal players (Brusco 1986) due to the challenge of having to gather a critical mass of actors for any direction of adaptation activity.

Building on the notion of the nature of clusters advanced in the existing literature, this study proceeds by investigating the role of cluster-level factors for the outcome of agent's self-organisation processes following an external change event. It does so for two reasons. First, the issue of architectural influences in clusters on the quality of adaptation by their agents is less explored than the notion of inertia at the level of individual organisations. Second, with a set of intriguing empirical findings at hand, it seems quite natural to attempt an investigation of the roles of division of labour and the mode of co-ordination for adaptation in clusters through a general theoretic model. In doing so, the study wants to derive clearer causalities than those obtainable from contextual case-study evidence. Finding and explaining differences in adaptive performance of clusters as a result of differing degrees of division of labour or modes of co-ordination could then contribute to an explanation of why specific empirical regularities can be observed. For instance, a greater adaptive performance of clusters with more hierarchical co-ordination structures could act as a justification for the aforementioned developments in many Italian districts.

1.2 Goals and contribution of the study

Against the background of the problems found in the many old industrial areas worldwide, this study investigates how agents in clusters can adapt to change in order to ensure the cluster's survival over time. As was outlined in the previous section and will be emphasised in more detail later (Chap. 4), cluster survival hinges on a number of aspects. At the micro-level, the success of adaptation depends on the quality of adjustments pursued by individual cluster organisations. Due to the nature of clusters, these are however constrained by architectural factors (agglomeration externalities and local culture). Both aspects together thus determine the outcome of the adaptation by self-organisation in clusters that directly follows a local bifurcation in the latter's development. This outcome in itself sets the stage for subsequent adaptation by arbitrage in the cluster as well as its host area. As a result, it has both immediate and long-term consequences for the development of the cluster as well as that of the greater region.

In this second, more long-term perspective of adaptation by arbitrage, additional aspects come into play. Within the greater environment of the cluster, it competes with other clusters or individual firms in the industry or related sectors. Alongside the outcomes of adaptation by self-organisation and arbitrage at the level of any cluster, this aspect of competition with other clusters or firms implies that the relative performance of the cluster following a change event as compared to these other entities will matter for survival. As a result, cluster adaptation and survival depends on three aspects: The outcomes of adaptation by self-organisation and arbitrage as well as the relative performance of the cluster with respect to other competing entities in its industry.

This study focuses on one of the three factors by investigating the causes underlying successful adaptation by self-organisation. It does so by analysing the role of architectural constraints on agent adjustment because in this context, the very advantages of clusters (agglomeration externalities and local culture) can turn into liabilities. Agglomeration externalities mean that changes in activity by agents impact on the outcomes of actions undertaken by others. The existence of local rules of the game can constrain the availability of adjustment measures for individual organisations if some options for adoption would entail defection and risk of punishment by other agents. While the cluster architecture thus influences the activities of its agents', the cluster itself has no formal authority over the decisions of actors. An interesting question in this context therefore regards, whether there can be any architectural factors that align individual activities with better collective outcomes.

Bearing the empirical examples of old industrial areas as well as the different responses of clusters in mind, it would be interesting to see for which cluster architectures adaptation by self-organisation can be successful. Unfortunately, the issue of challenges to and adaptation in clusters has been neglected in much of the existing literature. At the risk of oversimplification, one could say that in tendency, *theories dealing with dynamics do not deal with clusters while those investigating clusters do not concern themselves sufficiently with dynamics.* The models analysing the dynamics of the spatial distribution of industries and thereby clusters (Chap. 2) investigate the effect of the latter on macro-level factors such as growth, national welfare or trade flows. As a result, they do not concern themselves with the fate of individual areas affected by a shift in industry distribution. Those concepts having the cluster as their unit of analysis (Chap. 3) in turn do not sufficiently take the issue of dynamics into account. By focussing on the advantages from clustering they do not assess why, when and how these benefits can turn into liabilities due to changes in the area or the area's industry.

The empirical literature on cluster adaptation to date lacks conclusive analysis with respect to the factors underlying success or failure. This aspect stems from two sides. First, many contributions list a set of factors as relevant for adaptation (internationalisation of and co-ordination between local agents as well as extent of division of labour) but leave the underlying causalities relatively implicit. Second, current evidence on clusters, change and adaptation mainly lies with case study findings where clusters that were characterised by one or several of the aforementioned factors have been successful in surviving change. Due to the highly contextual evidence provided by each case, the observed phenomena can have multiple possible causes based on the specific circumstances. As a result, causalities are difficult to extract and the findings do not readily lend themselves to generalisation. The fact that surviving clusters are often characterised by a certain division of labour, some degree of internationalisation of their agents as well as specific ways of co-ordination between firms alone cannot explain why any of these factors could contribute to the success of their agents in adaptation. In addition, when comparing different empirical studies, a trade-off in the effect of the different factors can be found. For instance, some cases find that clusters structured as networks of small and medium sized enterprises (SMEs) are more adaptive than those with large, integrated firms (Saxenian 1994) while others argue that the former structure is a liability in adaptation due to the need to convince a critical mass of actors to make any necessary changes (Brusco 1986).

This study investigates the question of whether and when cluster architectures can help agents adapt to change events. It does so by deriving causalities with respect to two factors found to matter in empirical work: The extent of division of labour and the mode of co-ordination between cluster agents. The analysis uses complexity theory to provide a formal model of ideal-typical clusters, which differ in their 'architecture' according to these two factors. By obtaining and comparing simulation results on the adaptive performance of clusters with *differing and given architectures* regarding their degree of division of labour and co-ordination between agents, it aims at clarifying the relationship between these architectural features and adaptive performance more clearly than it has been achieved so far. The model will do so by shedding light on the causalities (how?) underlying any difference in adaptive performance of clusters with higher or lower degrees of division of labour and distinct modes of co-ordination. In doing so, it implicitly also indicates trade-offs (how much?) and interdependencies (which combinations?) that lead to a positive influence of division of labour and co-ordination on cluster adaptiveness. The approach adopted here thus falls into the category of *qualitative simulation modelling* (Valente 2005), i.e. it is meant to find *explanations* for stylised empirical phenomena rather than replicating specific past developments or deriving realistic parameter values for an existing theoretic model.[5]

In doing so, the study aims at three contributions to the literature in the field. First, it provides more conclusive analysis of the aforementioned causalities, trade-offs and interdependencies between the factors favouring adaptiveness according to the theoretic and empirical literature. In doing so, the model constitutes a first step towards an explanation of some striking regularities found in case studies of cluster's adaptation to change (Cainelli and Zoboli 2004b; Boschma and

---

[5] The latter approaches are prominent in much of the literature developing and employing simulation models. A replication of developments is attempted in *history-friendly* models. Other simulation approaches rely on *data fitting* between simulation and empirical results in an attempt to explain one observed development or in order to derive realistic parameter settings for an existing theoretic model.

Lambooy 2002; Lombardi 2003; Maggioni 2004; Saxenian 1994). Second, the identification of very 'adaptive' cluster architectures opens up a number of possible avenues for further research. Empirically, one could move towards comparative case study research of two or more clusters in an industry to see if the hypotheses advanced by the model hold. In this context, model results could also be used to assist policy initiatives by shifting some attention from spending resources in the very uncertain venture of *"creating the next Silicon Valley"* towards the project of *"keeping one's Silicon Valleys"*. Theoretically, a next step would involve the development of more agent-based models able to account for how clusters could move from one kind of architecture to another. Finally, it is hoped that this study will contribute to a more dynamic perspective in Marshallian agglomeration theory where this angle is currently missing.

1.3 Course of the study

This *first part* of the study has clarified the *why* of the study to be conducted here, i.e. the unit of analysis, research question as well as the latter's underlying rationale. The *second part* will address the *why not*, i.e. it outlines the limitations of the existing literature with respect to the research question pursued here. Reviewing concepts describing cluster development as a result of dynamics in the spatial distribution of industries (Chap. 2) yields that the latter investigate the macro-level impact of changes in industry distribution and do not concern themselves with the fate of individual clusters. The methodology and research questions underlying these theories have lead to the construction of models that are deterministic with respect to the behaviour of industries' spatial distributions (and thereby any clusters resulting from the former) after parameter changes. As a consequence, the models do not allow for cluster adaptation to an adverse change event and can thereby not account for the differential responses found in the empirical reality. Turning towards concepts with a focus on the cluster itself (Chap. 3) reveals that the determinism of the previous models is attributable to their limited account of proximity benefits. If the latter encompass effects beyond mere cost considerations (technological externalities), the aforementioned determinism breaks down as some externalities may have a non-linear effect. However, this Marshallian cluster literature is focussed on a static analysis of how agglomeration externalities make a cluster successful at a specific point in time.

Combining the understanding on what makes clusters successful advanced in the Marshallian literature with the idea that the spatial distribution of industries underlying the existence of clusters is anything but unchangeable then leads to *part III*. It addresses *how* the issue of change and adaptation can be analysed from a general theoretic perspective. Thanks to the nature of clusters as constructs of independent agents whose activities shape and are influenced by cluster level factors (Chap. 4), adaptation becomes a less than straightforward process of decentralised problem solving by local agents under interdependence at the cluster level. Nonetheless, empirical research on surviving clusters reveals a number of aspects that are repeatedly stated as important for the outcome of adaptation processes. In order to move beyond case study limitations, the study sets out to develop a modelling framework able to investigate the causalities underlying the stylised empirical observations. It shows that such a model can be derived if clusters are understood as a variant of complex adaptive systems (Chap. 5). Moving towards a complexity perspective on clusters, this part then argues that the N/K framework is particularly suited to the task due to its ability to account for the nature of clusters as well as the stylised empirical phenomena.

Part IV then outlines *what* was done in order to derive causalities for the observed empirical phenomena. Understanding clusters as adaptive systems of interdependent agent groups, it derives the roles of division of labour and mode of coordination between cluster agents in steering individual activities towards good collective outcomes. Reviewing the general properties of complex, co-evolving systems in the N/K framework (Chap. 6) assists in deriving behaviourally neutral choices for parameters that will not be investigated in the simulation model. Moreover, it assists in understanding the roles of division of labour and coordination for cluster adaptation within the chosen methodology. The model underlying the simulations is then introduced in Chap. 7. It views the vertical division of labour in a cluster as a factor influencing inter-agent interdependence and the mode of co-ordination as a determinant of the goals underlying agent strategy selection. While both aspects evolve throughout the development of the cluster, at the time the change event occurs, they constitute short-term fixed limitations for agent activities. Based on the insights on N/K system behaviour, both aspects will play a role in the system's (cluster's) ability to recover from a change event, i.e. on its ability to find good (fit) new solutions.

Part V presents the findings and conclusions of the study. In a way, it addresses the *so what* question by outlining how and why division of labour and coordination influence cluster adaptability as well as the kinds of insights obtained by this exercise. The results reported in Chap. 8 show robust outcomes indicating an optimum degree of division of labour. In addition, more collective modes of co-ordination increase stability and adaptive performance by gearing agent strategy selection towards fewer modifications that are better from the perspective of the cluster as a whole. Only in very interdependent clusters with central dominant actors can the latter convey an added stability benefit. However, very collective modes of co-ordination in absence of a formal enforcement tool (like uneven distributions of power in the cluster) can be destabilised through invasion of egoistic agents as each group's fitness values show a prisoner's dilemma payoff structure. Chap. 9 summarises the results obtained and the insights gained from them with respect to the stylised empirical phenomena involving the Silicon Valley – Boston 128 comparison on the one and the experience in many Italian Industrial Districts on the other hand. While the model can be shown to outline causalities on why and how division of labour and co-ordination help cluster agents adapt to change in a more conclusive manner, its limitations invite a number of future research avenues regarding empirical tests of its hypotheses as well as extensions focussing on the determinants of agent behaviour. While the comparative dynamics conducted here to single out the roles of division of labour and co-ordination are only a first step towards an understanding of cluster dynamics, the study does show that a Marshallian perspective on the nature of clusters can be reconciled with dynamics when adopting a complexity perspective. It would be hoped that future research will explore this open path to a greater extent than it has been accomplished here.

Part II Literature review – The benefits of co-location

> *When an inventor in Silicon Valley opens his garage door to show off his latest idea, he has 50 per cent of the world market in front of him. When an inventor in Finland lifts his garage door, he faces 3 feet of snow.*

Jorma U. Nieminen, CEO Nokia[6]

The following chapters address the question of why the issue of cluster adaptation to change cannot be covered with the tools provided in the existing literature. It finds that while the argument for the existence of clusters always lies with agglomeration externalities, their respective treatment differs between the concepts introduced in chapters 2 and 3. This has important repercussions for the issue of clusters, change and adaptation: Those theories investigating cluster dynamics as a consequence of the developments in the spatial distribution of industries (chapter 2) arrive at deterministic models where changes in parameters result in an inevitable spatial concentration of industries equalising cluster growth or a spatial dispersion of firms that implies decline of existing clusters. Agents in a cluster have no means of changing these outcomes through adaptation.

The deterministic outcomes in the models provided are attributable to their limitation of proximity benefits to pecuniary effects influencing the benefits and costs of being in an area. The concepts describing the success of clusters as attributable to both pecuniary and technological externalities (chapter 3) in turn are more involved with studying how both aspects work to make a specific cluster successful at a given point in time. As a result, one could say (at the risk of oversimplification) that those concepts investigating spatial dynamics do not cover the topic (adaptation). Those concepts that would be able to include adaptation are not involved with a study of dynamics. The literature review will show that accepting the notion of clusters as spatial constructs exhibiting both pecuniary *and* technological externalities between agent activities means that their adaptation to change is no longer a deterministic process. Due to the possible non-linear effects of some technological externalities, agent adaptation in such clusters can only be studied by advancing another perspective on the phenomenon which is based on new findings in complexity science.

---

[6] Quoted in: van Tulder 1988, p. 169.

2 Stability and change: Driving cluster development

As has been advocated elsewhere, a full 'theory' of the cluster would have to provide an answer to the questions of 'what, how, why as well as when, where and who'. Its main focus, however, should lie in identifying the causalities, i.e. in addressing the 'why' question (Maskell and Kebir 2004). In the context of clusters, this comprises three aspects: 'Why' do they emerge, endure for a time and why do they become exhausted? The existing literature investigating the drivers of cluster development finds them in the duality of stability and change in the spatial distribution of industries. Clusters emerge due to a series of events that leads to an initial concentration of firms in an area. Their endurance is then attributable to the stabilising effect of positive and negative agglomeration externalities (Richardson 1993). As a result, the local firm population grows up to a threshold level often referred to as the area's 'carrying capacity'. The exhaustion of the cluster is then brought about by changes in the effect of agglomeration externalities (usually by altering model parameters). The outcome of this event with respect to the cluster is a deterministic one, i.e. there is no allowance for adaptation beyond changes in the spatial distribution of the industry. In the face of adverse local bifurcations, exhaustion is thus inevitable.

The overreaching theme in this chapter lies with the notion that the development of a cluster is determined by the location choice of firms, which is in turn influenced by the attractiveness of a locality. The latter develops over time due to the effects of agglomeration externalities and outside events. Of course, there are many models of regional (cluster) development in the literature. The selection of the approaches reviewed here is based on their proximity to the concepts in the Marshallian tradition (Chap. 3): Both lines of thought view the same group of agglomeration externalities as the source of endurance for clusters.

A first concept of cluster development (2.1) links the evolution of a cluster to the dynamics of its industry. Arguing that agglomeration externalities are most important in the early stages of industry development (due to uncertainty and new technological knowledge), a cluster is viewed as a product of historical accident leading to an initial concentration of firms in an area. The latter then out-compete isolated entities due to agglomeration externalities in the transfer of (tacit) technological knowledge and the co-ordination of a division of labour. As the industry matures, technology becomes standardised and production more integrated, thus reducing the benefits from proximity. At this stage, the cluster's exhaustion can be brought about by the emergence of new competitors from lower cost locations or through the birth of a new industry replacing the existing one. However, the generality of industry dynamics underlying the concept has been questioned. Moreover, the empirical evidence of Sect. 1.1.3 suggests that cluster exhaustion can be brought about by smaller events than the exhaustion of entire industries. As a consequence, more general insight into cluster dynamics can be obtained if approaches focus on the developments found at the local level and allow for more varied sources of change.

Sect. 2.2 introduces these more general theories of cluster development where endurance is modelled (in different ways) as a result of agglomeration externalities and where change is represented as any event altering model parameters.[7] The first model argues that positive agglomeration externalities (agglomeration economies) lead to symbiotic processes where the growth of the local firm population reinforces itself. This growth is however constrained by location-intrinsic factors like infrastructure acting as habitat constraints. The critical mass of local firms for establishing a sufficient strength of the symbiotic processes is introduced by historical accident. The cluster then grows through 'local symbiosis' up to the area's carrying capacity. Exhaustion is in turn brought about by another external event changing the strength of the symbiotic processes below a threshold value, thereby resulting in a dispersion of the industry. A second set of models found in the New Economic Geography (NEG) relates cluster evolution to locational profitability as the key factor underlying an industry's spatial distribution (2.3). Small differences between areas can – when coupled with suitable model parameters – allow for the emergence of a process of circular causation where an industry's spatial concentration reinforces itself. The contrasting forces of positive and negative pecuniary agglomeration externalities then govern the extent of the concentration process, which stabilises the cluster's size at a specific level (endurance). This equilibrium is challenged by events changing model parameters, allowing for an increase in cluster size or a dispersion of the industry (cluster exhaustion).

A third set of models argues that both an area's physical endowments and the size of its firm population (through agglomeration externalities) contribute to its attractiveness. Firms differing in their preferences for area-intrinsic factors choose the most attractive site for their operations. By allowing for a random order in the sequence of firm entry, historical accident (in terms of the location choice of early entrants) leads to a path dependent process yielding different spatial distributions of the industry. Areas ending up as hosts to industry clusters once more grow to their carrying capacity. An exhaustion of these clusters can be brought about by external events changing the benefits conveyed by area-intrinsic factors or those related to agglomeration externalities. Depending on the strength and direction of their impact (increase/ decrease of benefits), such events can lead to a redistribution of the industry, implying exhaustion of existing clusters.

Two aspects will carry from this literature review throughout the remainder of the thesis. The first is an adoption of the notion that a duality of change and stability drives cluster development. The second regards the view of agglomeration externalities as the source of cluster stability. Both aspects have specific implications

---

[7] To keep a flow in the argument outlined in this chapter, formal representations of the different model have been omitted and are provided in the respective appendices (1-3).

for the further course of the study. For one, at the time of a change event carrying the potential of cluster exhaustion, the latter is characterised by a multitude of agglomeration externalities producing interdependence between its agents. As a consequence, a study of the factors favouring or hampering the *adaptation* of cluster agents in the face of local bifurcations requires a different modelling approach able to account for the effects of both agent activities and the influence of agglomeration externalities.

2.1 Novelty, uncertainty and transaction costs: Industry-location life cycles

A first avenue to investigating cluster development as determined by the spatial distribution of an industry regards an analysis of the developments in the latter. This approach argues that during what could be called a "industry-location" (IL) life cycle, the technological foundations of an industry evolve and change the nature of its operations (Klepper 1996). This in turn leads to a change in the benefits from firm co-location and thereby the spatial distribution of the industry. Clustering is advantageous in early stages of the IL life cycle: First, while technological knowledge is new, it lacks standards for its codification (e.g. definitions and specifications of terms). In consequence, access to or transfer of this *tacit technological knowledge* (Pavitt 1987) requires repeated personal contact with the people embodying it, which is more easily obtained in proximity. This induces a co-location of firms in the emerging industry with sources of technological knowledge (research institutions or other firms). Second, the uncertainty and small firm size in the early stages of the cycle favour a division of labour between firms regarding production in order to avoid overcapacities (Audretsch and Feldman 1995; Klimenko 2004). Such a division of labour requires co-ordination between the different parties involved, which is facilitated by their co-location (Storper 1995).

In an early stage of the industry life cycle, co-located firms would thus benefit from transaction cost savings. If historical accident leads to a concentration of firms, they out-compete isolated entities and a cluster emerges. As the industry matures, a dominant product design is developed and increasingly adopted by users. Growing demand alongside more fixed product characteristics then allow for a greater integration of production (less division of labour) and firm growth. In addition, the industry's technology becomes more standardised. As a result, the importance of proximity for knowledge transfer and inter-firm co-ordination declines. The second stage of the cycle would therefore see a consolidation of firm numbers and a growth of the cluster as a whole. This increases competition in input, infrastructure and end-markets, bringing up costs and decreasing locally obtainable product prices, i.e. negative agglomeration externalities start to materialise, limiting cluster growth.

In the third stage, the product becomes even more standardised making cost rather than performance the driving force in competition. Taken together with the smaller role of clustering benefits (transaction cost savings), the costs to being in the cluster (negative externalities) come to be overwhelming. This opens up possibilities for a dispersion of industrial activity (Storper 1997), which is brought about by an external event in the guise of competition from lower cost areas or the emergence of a new industry. In case of competition from lower cost producers, local firms can shift (parts of) the production to cheaper areas, be replaced by firms from other locations or move to upper market segments to evade price competition (Tichy 2001). All three developments imply a partial or full exhaustion of the cluster as local economic activity is diminished by relocation or shrinking markets. The emergence of a new industry in turn leads to a full exhaustion of the cluster. As the discoveries leading to new products are usually made by industry outsiders that do not necessarily locate in the high cost environment of an existing cluster (Davelaar and Nijkamp 1990), new industries tend to emerge in new locations. If in addition, no linkage between the new and the existing sector exists (in the sense that incumbent's experience is not relevant for the new industry; Swann 1996), 'old' firms will tend to stick to their current technology. With time, their products are replaced by the new industry and the old industry disappears from all clusters (McGahan 2004).

Consequently, the IL life cycle argues that clusters are created in the early stages of industry development by historical accident. Transactions cost advantages in the transfer of tacit technological knowledge and the co-ordination of a division of labour sustain the cluster for a limited time. Its extinction is brought about by developments in the industry leading to cost competition and a spatial dispersion of firms or to the replacement of an existing industry by a new one.

The first set of objections advanced against the IL life cycle regards the generality of industry dynamics ascertained. It is likely that only some industries grow towards product standardisation, vertical integration, mass production and cost competition (Klepper 1997; Acemoglu et al. 2004), which leads to a pattern of agglomeration and dispersion in their spatial distribution. This could explain that clusters are found in both new and relatively mature industries (e.g. Italian Industrial Districts; Pyke et al. 1990). In addition, neither the possibility of reinventing maturing industries with new products or processes nor the notion of technology convergence between old and new sectors (Swann 1996) are accounted for.

A second concern regards the treatment of agglomeration externalities. As will be elaborated in more detail in the following chapter, these extend well beyond transaction cost savings. In consequence, other agglomeration externalities can work to stabilise the cluster in later phases of industry development even if tacit technological knowledge or a division of labour loses in importance (Maskell 2001).[8] If industry developments cannot be generalised sufficiently to explain cluster dynamics, their study may be better addressed at a more general level

---

[8] This of course does not mean that agglomeration economies cannot be altered or eliminated by industry developments. The extent to which they are present in different areas might also explain why an industry can display different "temporal, spatial, and organizational patterns of development" (Storper 1997, p. 60), i.e. why it agglomerates in small firm clusters in some regions and is dominated by large, dispersed corporations in others (Rabellotti 1995).

through an investigation of the dynamics of the attractiveness of locations. Industrial developments - among other things – will play a major role in this context. The different models relating agglomeration development to changes in location attractiveness and firm location choice (Sects. 2.2-2.4) achieve this.

2.2 Symbiosis, habitat and external events: Dynamics of firm populations

The first of the 'general' models of cluster development (Brenner 2000, 2001 and 2004) views it as a result of internal processes and external events affecting the industry's spatial distribution. The internal processes consist of the role of habitat constraints posed by location-inherent factors that are difficult to change in the short run (e.g. area size or resource endowment) as well as the effects of a growing local firm population (agglomeration externalities). External events in turn are all factors impinging on the strength of agglomeration economies or the number of local firms irrespective of the activities of the latter (e.g. policy measures to increase personnel training effort or growing demand due to industry growth).

The development of the cluster is then measured by changes in the local firm population (represented as a composite of firm and employee numbers). For every set of external conditions (e.g. market size) and inherited local characteristics, a region exhibits a maximum cluster size. This 'carrying capacity' also depends on the extent of positive agglomeration externalities (agglomeration economies). The latter are conceptualised as mutual profits to firms from their co-location (Brenner 2000), which result from:

- Direct effects (inter-firm co-operation, local information spillovers or venture capital provision by incumbents) between existing firms in an industry;
- Indirect effects (accumulation of human capital, industry-tailored infrastructure or political support), where firm activities improve production conditions for existing agents and facilitate new firm entry; and
- Cross-industry effects where a supplying industry relies on the growth of its customer sector.[9]

The existence of agglomeration externalities then introduces a symbiotic relationship between the number of local firms in the industry and the size of that population, local conditions as well as the number of firms in the supplying industry. Unlike the dynamics of symbiosis in the biological context, the growth rate of a cluster is not linear and symbiotic processes are uneven.[10] Moreover, symbiosis is not linearly dependent on population states: Due to congestion effects, firms do not always benefit more from agglomeration economies the greater their number.

---

[9] Mathematically, all three types of effects are treated identically as symbiotic relations. An extension to the model includes the case of complementarities between industries s and f where the cross-sectoral symbiosis becomes reciprocal (not considered here).

[10] For the formal representation, please consult appendix 1.

To account for this, the symbiotic relationships representing agglomeration economies in the model only emerge after a critical mass of local firms is passed. Within a certain range of cluster size, they remain constant.[11] Congestion effects are accounted for by arguing that once the size of the cluster exceeds a threshold value, they come to outweigh agglomeration economies and a growth constraint sets in. The maximum size of the cluster (the area's carrying capacity) then depends on inherited (e.g. size or resource endowments) as well as external conditions (e.g. market demand).

For a given set of model parameters, an agglomeration of industry activity becomes possible due to the existence of two stable equilibria. One represents the 'dispersed' state where the local firm population remains low. The second is characterised by 'agglomeration', i.e. a large firm population in the area. The existence of suitable parameters regarding the initial values of external conditions, inherited local characteristics and agglomeration economies is justified by historical accident. The forces inducing a shift between these two equilibria then shape cluster development. In its first stage, a change in external conditions increases the strength of symbiotic processes. This is achieved directly or indirectly, i.e. by efforts increasing the strength of agglomeration economies or by activities enabling greater numbers of local companies (emergence). Examples could be an increase in demand in a new industry or policy initiatives supporting start-up activity (Brenner 2003). As a result, the area 'moves' to the second equilibrium of heightened local economic activity, i.e. a cluster emerges. The latter is then stabilised by agglomeration economies against small changes in external conditions (endurance). If changes in the external conditions however come to exceed a threshold value, symbiosis is rendered non self-enforcing (exhaustion). This change can again be brought about directly (reducing the strength of an agglomeration economy) or indirectly (through a reduction in numbers of firms). In both cases, the cluster vanishes as the local firm population returns to the low equilibrium value. The symbiosis resulting from agglomeration economies becomes latent and might – in the wake of another exogenous event – lead to the emergence of a new cluster on the 'ruins' of an old one.

Summing up, the model asserts that clusters emerge due to symbiosis in a local firm population resulting from agglomeration economies. To obtain a sufficient strength in these symbiotic relations, an outside event is needed. Once the industry has agglomerated in a region, it is stabilised by agglomeration economies until external conditions change sufficiently to revert it to the dispersed state. Put differently, the development of a cluster is described as the result of changes in external conditions with a period of stability (cluster existence) between them. Again, a cluster is only sustainable for a limited period of time.

A first criticism to be advanced regards the model's optimistic perspective on cluster renewal where it takes but another exogenous event for a new cluster to emerge on the ruins of previous ones. Empirical evidence of regions undergoing

---

[11] This constitutes an important difference between the concepts presented in this section and those in Sect. 2.3. There, the strength of agglomeration economies is a function of the number of local firms.

structural change highlights that industrial substitution is more difficult to achieve than suggested here (Braunerhjelm et al. 2000, Dalum et al. 2002, Grabher 1993, Schamp 2000, Tiberi Vipraio 1996). Moreover, the symbiotic model faces the problem that it describes the development of an isolated cluster. In reality, firms will have different regions to choose as the site of their activity and the outcome of this location choice determines if any area comes to host an agglomeration of firms. Thus, central aspects like the competition between different clusters are not included in the model. This issue is addressed in the poly-regional models presented in the following two sections.

2.3 Externalities and trade costs: The New Economic Geography

It has been mentioned in the introductory section that the spatial distribution of economic activity was long left outside the focus of mainstream economics. This was mainly due to the predominance of general equilibrium models where the formal representation required very restrictive assumptions: Until the late 1970s, trade theory was based on models with perfect competition, constant returns to scale in production as well as absence of transport costs. In such a setting, clusters could not emerge, as constant returns to scale meant no advantages to concentrating production in a plant or an area. A lack of transport costs did not allow for any benefits from proximity to markets or inputs. Consequently, any spatial concentration of productive activity had to be credited to 'natural' differences between regions, e.g. regarding resource endowments or population density.

Geographic concerns only re-entered mainstream economics when new formal tools made it possible to deal with transport cost and increasing returns to scale.[12] Increasing returns to scale were addressed within the framework of monopolistic competition (originally developed by Chamberlin 1933 and formalised by Dixit and Stiglitz 1977). It argues that consumers have a preference for variety rather than quantity in the sense that they derive a higher utility from the consumption of smaller quantities of more varieties than from consuming large quantities of fewer varieties. As a consequence, distinct product varieties enter symmetrically into demand. Since producers attempt to maximise profits, only one firm produces each variety. The 'monopoly' in this variety is however constrained by other manufacturers offering substitute products.

---

[12] Initially, these tools were used in trade (Helpman and Krugman 1985) and growth theory (Grossman and Helpman 1991). For NEG models, see among many others Bischi et al. 2003 Fujita et al. 1999; Fujita and Thisse 1996; Krugman 1991a, 1991b; Nicoud 2004; Ottaviano and Thisse 2002; Puga 1998; Roos 2002b; Venables 1996, 1998. Appendix 2 introduces the 'core' NEG model.

Transport costs are treated as 'iceberg' costs, i.e. part of the goods shipped to distant markets 'melts' away in the process.[13] The amount of product loss grows with distance, implying that only a reduced amount of the original shipment reaches its destination. This increases the price of non-local varieties. The total costs of accessing non-local markets then depend on transportation costs as well as trade barriers. The composite of both aspects is referred to as *trade costs*.

The NEG models analyse the dynamics of the spatial distribution of an industry as a result of firm location choice for a two-region (country), two sector (agriculture and manufacturing) situation. Only one of the sectors (manufacturing) is subject to increasing returns to scale and trade costs, which allow for different spatial distributions. Increasing returns and costly transportation imply that manufacturing firms will concentrate production of their variety in one site that should preferably be near a large market *"to avoid trade costs in a larger fraction of [...] sales"* (Ottaviano and Puga 1997, p. 5). The markets for manufacturing products lie in the agricultural sector as well as manufacturing itself. The latter link allows for a process of *circular causation* where a spatial concentration of manufacturing firms gives rise to further concentration by increasing the size of the local market (Krugman 1991b, p. 486).

The positive market effect is however constrained by congestion phenomena emerging if manufacturing firms concentrate in one region and service the other through trade. While greater local demand allows firms to capitalise on scale economies, a growing firm population also increases the cost of production factors and inputs in the region. Moreover, increased end-market competition reduces local product prices thereby lowering the value of a large share of firms' sales. The more costly transport to the other region and the lower scale economies, the sooner these effects will be felt in firm profitability. Both effects (scale economies and congestion cost) thereby drive the spatial distribution of the manufacturing industry towards equilibrium where the profitability of firms in both regions is equalised.

In the setting described so far, any agglomeration of manufacturing firms would be exclusively due to area-intrinsic differences, e.g. regarding size (population). A clustering of manufacturing beyond area-intrinsic factors is made possible through the inclusion of pecuniary (Scitovski 1954) agglomeration economies in the model. In Krugman 1991b, the latter stem from labour mobility in the manufacturing sector: As product prices reflect the transport cost for non-local goods, regions with a larger share of the manufacturing sector face lower price levels. This in turn increases real wages for workers and induces a migration of manufac-

---

[13] Paul Samuelson 1954 is reputed to have coined the term *Iceberg transport cost*. However, the idea of product *consumption* during transport goes back much further. Johann Heinrich von Thünen 1826 introduced it in a more literal sense arguing that a horse (and the people) transporting wheat from the countryside to the city market eat *"a fraction of it en route"* (Jones and Kierzkowski 2004, p. 168). That fraction obviously increases with the distance between both sites, thus making transportation unsustainable at some point. *"Unter diesen Verhältnissen ist also der Transport des Korns auf 50 Meilen unmöglich, weil die ganze Ladung oder deren Werth auf der Hin- und Zurückreise von den Pferden und den dabei angestellten Menschen verzehrt wird"* (von Thünen 1826, p. 8).

turing labour towards the region. With growing local labour supply, nominal wages decrease, which in turn attracts more manufacturing firms into the area and further reduces product prices, enabling a process of circular causation.

Later work in the NEG framework has introduced other sources of agglomeration economies including economies of scale in supplier (Krugman and Venables 1996 or Venables 1996, 1998) or servicing industries (Baldwin 1999) inducing a positive externality on input and service costs. The extent of clustering is then constrained by the relative profitability of manufacturing firms in either of the two regions.[14] Once profits between both groups are equalised, relocation no longer pays and a stable equilibrium is obtained. The latter can be found with a homogeneous distribution of the industry between both areas, differing degrees of spatial concentration or even a monocentric outcome where one of the two areas appropriates the entire manufacturing sector. In other words, "*[t]he interesting finding is that, for finite positive trade costs, the core's share of world industry is larger than its share of world endowments*" (Ottaviano and Puga 1997, p. 4).

Again, the process of cumulative causation resulting in a clustering of manufacturing firms is constrained by the aforementioned congestion effects in factor, input and product markets and their impact on firm profitability. The latter is found to rely crucially on model parameters regarding the size of the manufacturing sector, the extent of scale economies and the level of trade cost: "*In an economy charaterized by high transportation costs, a small share of footloose manufacturing or low economies of scale, the distribution of manufacturing production will be determined by the distribution of the "primary stratum" of peasants. With lower transportation costs, a higher manufacturing share or stronger economies of scale, circular causation sets in, and manufacturing will concentrate in whichever region gets a head start*" (Krugman 1991b, p. 497). The bigger the manufacturing sector, the greater its possibility to generate demand for itself by concentrating spatially, thus making the spatial distribution of manufacturing increasingly independent from that of agriculture. Moreover, greater benefits from scale economies and lower costs of servicing the distant market decrease the 'cost' of increased factor, input and end-market competition in a cluster.[15] Regarding scale economies and trade costs, clustering is only possible for intermediate values. Too strong scale effects can always produce clustering of the sector whereas too low values mean that concentrating production in one region no longer pays. In that case, firms might begin to produce in both regions inducing a more dispersed distribution of the manufacturing sector (in analogy to Helpman and Krugman 1985). Prohibitive trade costs limit firms to servicing the local market meaning that congestion effects set in very quickly. Any agglomeration in this context would therefore once again depend on location-intrinsic factors such as the size of the local

---

[14] This reasoning is borrowed from location choice theory where profitability determines production sites (e.g. von Thünen 1826, Weber 1909, Lösch 1962 and Christaller 1966).

[15] The notion of congestion costs has later been included explicitly in the analysis. They emerge due to immobile labour (Puga 1998) or untradeable inputs (Helpman 1997). As has been argued elsewhere (Nicoud 2004), these differences in mechanism underlying agglomeration economies and congestion costs have not altered the mathematical structure of the core NEG model significantly.

market. If trade costs are very low, firm location choice becomes conditional on production cost implying that the negative input and factor-market effects of agglomeration would induce a spatial dispersion of manufacturing firms.

Historical accident initialises the dynamics of the NEG model, e.g. because workers are distributed homogeneously *within*, but not necessarily between regions. These small differences set off a circular causation process inducing a spatial concentration of manufacturing in one of the regions until profitability of firms in both areas is equalised or until one region appropriates the entire sector. This equilibrium can be challenged by changes in model parameters. The most frequently discussed case regards changes in trade costs: Increases in trade costs lead to a greater focus on the local market. In this context, congestion effects will materialise in larger firm populations, and therefore lead to a dispersion of manufacturing activity. In contrast, a decrease in trade costs reduces local product market competition as distant markets become increasingly accessible. This introduces a possibility for further clustering of manufacturing firms.[16] Any change in trade costs therefore produces a deterministic change in the spatial distribution of the industry towards increased agglomeration or dispersion. The models of the NEG thus show that in the presence of increasing returns to scale at the firm level, small causes (such as small size differentials between areas) can have major effects regarding the geographic distribution of industries. The relative influence of agglomerative and deglomerative tendencies governing the spatial distribution of the manufacturing industry is thus mediated by trade costs.

In NEG models, the development of a cluster is therefore again subject to the duality of stability and change. The assumption of small initial differences constitutes a first external event leading to the emergence of a cluster (given suitable parameter values). Agglomeration economies and congestion cost then induce cluster growth up to the area's 'carrying capacity'. Unless another change alters model parameters (e.g. regarding trade costs), this cluster prevails over time.

Since its introduction, the NEG has attracted both acclaim and criticism.[17] The latter has stemmed from different sources. One regards the assumptions underlying the models, where the strife towards formal representations has lead to assumptions that are both 'artificial' even to proponents of the NEG (Fujita et al. 1999) as well as sometimes 'arbitrary' in the sense that they follow more the need to generate numerical solutions than actual economic reasoning (Scott 2004, p.

---

[16] See also Marshall 1920, p. 273: "*Speaking generally we must say that a lowering of tariffs, or of freights for the transport of goods, tends to make each locality buy more largely from a distance, what it requires; and thus tends to concentrate particular industries in special localities.*" Should trade costs fall below a threshold value, the effects of increased input and factor costs in clusters again come to outweigh the benefits, thus leading to a dispersion of manufacturing.

[17] Criticism has come especially from geographers: Martin and Sunley 1996 and Martin 1999 have serious concerns regarding its 'novelty' and 'geographic content'. More balanced accounts on the contribution of the NEG are found in Scott 2004; Neary 2001 or Schmutzler 1999.

486).[18] Furthermore, the exclusive focus on pecuniary externalities (cost reductions from agglomeration) has been a source of objection. While Krugman 1991b, p. 485 justifies this focus as a means to *make the analysis much more concrete than if [...] external economies [were allowed] to arise in some invisible form.*", other authors argue that it severely limits the contribution of the NEG to the understanding of 'real places' and their development: "*despite the fact that Krugman and his co-workers make frequent reference to Marshall, the model actually gives short shrift to any meaningfully Mashallian approach to regional development and agglomeration*" (Scott 2004, p. 487).[19] The limited treatment of agglomeration economies in the NEG has important consequences for the dynamics of industry distribution in the sense that it enables determinism in the response of regional clusters to changes in parameter conditions, not allowing for any adaptive measures other than a change in industry distribution (see also Chaps. 3 and 4).

2.4 Increasing returns and firm location choice: Spatial path dependence

The idea of circular causation in the concentration of industries is also taken up in models describing cluster development as a path dependent process. Originally developed to study the diffusion and adoption of technological standards (Arthur 1989a; David 1985), the notion of increasing returns was soon extended to the case of regional development. Arguing that a spatial concentration of firms in an area improves its attractiveness through positive agglomeration externalities (e.g. regarding labour and input availability), an idea running through much of the work of the German School of regional economics (in particular Engländer 1926; Ritschl 1927 and Palander 1935) was formally accounted for: The spatial distribution of firms in an industry at any time is dependent on the previous one (Arthur 1990, p. 236).[20]

---

[18] Examples in this context include iceberg transport costs and Dixit-Stiglitz consumer preferences. The unrealistic assumptions of the NEG models have also limited their empirical testability. Notable exceptions include surveys on the effects of economic integration (eliminating tariffs as a part of trade costs between countries) on the geographic distribution of industries (e.g. Brülhart 1996; Davis and Weinstein 2003, 2002 or Hanson 1996b and historically Crafts and Mulatu 2004). For overviews of empirical work in the NEG framework see: Overman et al. 2001 or Head and Mayer 2003.

[19] Although the reasoning advanced in Krugman's core model resembles Marshall's idea of pooled labour markets (see Chap. 3), there is an important difference. In the former concept, pooled markets provide a better quality <u>and</u> a lower cost of labour due to cumulative effects of individual firm behaviour. In the NEG, pooled labour markets only reduce labour cost. The inclusion of more technological agglomeration externalities is however starting in the more recent NEG models (e.g. Combes 1997; Ottaviano and Thisse 2002 or Harrigan and Venables 2004).

[20] For evidence, see Head et al. 1995 or Rauch 1993.

Firms in the model decide on locating in one out of Q possible areas. They are well informed on the profitability of each location, selecting the area offering the highest benefits at the time of their choice. The benefits of each area depend on two factors. On the one hand, all areas exhibit location-intrinsic factors like raw materials or access to transportation networks. These factors convey a certain *geographic benefit* to firms, which differs according to firms' (randomly distributed) locational preferences. On the other hand, the benefits to locating in a site are influenced by the number of local firms as the latter generate *agglomeration benefits*, which are identical for all enterprises. The total benefits to locating in any area are then the sum of geographic and agglomeration benefits. They change over time as more and more firms enter the industry and alter their area's agglomeration benefits by choosing a location.

At the start of the process, all areas are empty meaning that geographic benefits are the chief cause of location choice. Firms with differing locational preferences enter in a random sequence and chose their location according to the highest profitability obtainable at the time of entry (Arthur 1989b). The latter depends on firms' locational preferences (geographic benefits) and the area's current firm population (agglomeration effects). Once firms are established in an area, they do not relocate. As the industry grows, locational shares of industry activity stabilise into equilibrium (see appendix 3). This process is shaped by the distribution of locational preferences between firms as well as 'historical accident' in the sequence of entrants. The former determines how many firms would choose a specific region based on its geographic benefits alone. This effect is however distorted by the existence of agglomeration benefits: If one area attracts a sufficient number of firms, the impact of agglomeration can bring its total benefits to a level exceeding that of all other areas. During the early phases of industry development geographic benefits and early entrants' locational preferences therefore determine whether or not, some regions can get ahead of others to appropriate a significant share of the industry.[21]

The spatial distribution of an industry thus depends on the heterogeneity of firms' locational preferences and the relationship between agglomeration benefits and local firm numbers. In absence of agglomeration effects, firm location choice depends exclusively on locational preferences and the resulting geographic benefits offered by each area. As the industry grows large, its distribution will stabilise according to the distribution of locational preferences: The more homogeneous the latter, the more spatially concentrated the industry. Greater heterogeneity would in turn lead to a greater spatial dispersion.[22]

---

[21] The model also provides an explanation of why areas neighbouring an agglomeration are often orphaned regarding industry activity. If neighbouring regions exhibit similar geographic characteristics, locations with more firms exhibit greater total profitability, as agglomeration benefits are larger. Consequently, new firms would locate there, creating an 'agglomeration shadow'.

[22] It is interesting to note that model allows for the agglomeration of an industry in the absence of any agglomeration economies. Examples would be sectors depending very strongly on certain geographic endowments, e.g. extracting industries relying on raw materials (Arthur 1990, p. 241).

If agglomeration effects are unbounded (no congestion effects), total locational benefits will always increase with the number of local firms. As a consequence, the profitability of one or very few areas can increase sufficiently for them to become the profit-maximising location for all firms. Which area(s) end up as the 'winning' regions in this context depends on geographic attractiveness (i.e. the share of firms in the industry exhibiting locational preferences for the area) and historic accident in firm entry (did enough firms with a preference for one area enter early enough to lead to a sustainable advantage from agglomeration benefits).

The most interesting case relates to a situation of bounded agglomeration economies, i.e. the notion that adding firms to the area increases total locational benefits but not without limits. This aspect can be incorporated in two ways. The first argues that once the local firm population grows beyond a threshold level, agglomeration benefits become zero, i.e. total locational benefits remain constant. The second way includes an explicit inclusion of congestion effects meaning that at some point in time, net locational benefits decrease with the number of firms as congestion 'costs' come to outweigh agglomeration benefits. In the first case of bounded, non-negative agglomeration economies (Maggioni 2002), one area can monopolise the industry if enough early entrants locate there and increase its total benefits sufficiently for all future entrants to prefer it over other regions – even if their own entry no longer increases locational profitability. At the same time, several areas (two or more) can come to share the industry provided early entrants prefer different areas and lead to a parallel growth in agglomeration benefits among these. What decides on the monocentric or polycentric outcome is again the heterogeneity of firms' locational tastes (implying that one or several regions can be attractive to early entrants), their sequence of entry (which firm 'types' come first) as well as the strength of agglomeration benefits (determining the 'lead' of initially chosen areas).

In the case of bounded, negative agglomeration economies, the location choice of firms depends on the benefits and the *costs* incurred, which are in turn influenced by 'geographic' and 'agglomeration' effects. Put differently, parts of local benefits and costs are "*unaffected by the number of incumbents*" while others "*depend on the number of incumbents*" (Maggioni 2002, p. 97). The geographic benefit and cost component is assumed to be constant over time, i.e. net locational benefits depend exclusively on the relationship between agglomeration economies and congestion costs and thereby the number of local firms. In this extension to Arthur's model, agglomeration economies are viewed as initially increasing with the size of the local firm population and decreasing once the cluster grows beyond a threshold level. Congestion costs in turn initially decrease as the local firm population grows and begin increasing after a certain cluster size. As a result, net locational benefits initially increase with the number of firms as the positive effects of agglomeration economies outweigh the constraint posed by congestion costs. Once a threshold cluster size is reached, net locational benefits decrease as the greater competition for local resources comes to outweigh agglomeration benefits. As a consequence, the cluster growth process stops at the carrying capacity where net locational benefits are equal to zero.

The spatial distribution of the industry then depends on the hypotheses regarding firm locational preferences and industry growth. If there is heterogeneity in firm locational preferences and industry growth tends towards the indefinite (Arthur 1989b), the inclusion of congestion costs leads to a more homogeneous distribution of the industry with more attractive regions 'filling up' first. If however firm locational preferences are homogeneous and industry growth is limited, the spatial distribution of firms is likely to result in a monocentric outcome with the region offering the highest geographic benefits ending up as the industry's centre (Maggioni 2004).

While simulations do highlight the key role of early events for the resulting spatial distribution of an industry (Brenner and Weigelt 2001),[23] a critique can be advanced against three hypotheses in the models of path dependence. The first regards the issue of unlimited industry growth. It is fairly intuitive that the number of firms in an industry cannot grow without limit for the sake of demand constraints and competition. In this case, the agglomeration process ceases at a specific point in time, giving both agglomerative and dispersal forces less time to operate. The final spatial distribution in this situation would on average tend to be more polycentric due to the randomness of firm entry. Second, the question of what happens to firms located in areas that get 'left behind' by the emerging clusters poses itself. These firms experience a lower locational profitability than their clustered counterparts, thus reducing their profitability as a whole. It can be argued that such firms would either have to go out of business or relocate. Either dynamics are however not accounted for.

Finally, the issue of cluster exhaustion is underrepresented.[24] Given the underlying hypotheses, a change in industry distribution can only be brought about if some development in the industry changes the benefits from geographical and/ or agglomeration related factors or if a new industry with different locational preferences and agglomeration benefits emerges. In both cases, a change in attractiveness of areas occurs with the now more attractive regions experiencing further entry by new firms or through relocation of existing ones.

---

[23] In the model, path dependence is shown for a two-industry, multi-regional framework. Simulations indicate a trend towards an equilibrium distribution of the industries. This equilibrium outcome is strongly shaped by early events in the simulation process, thereby supporting the notion of path dependency in an industry's spatial distribution.

[24] The models of path dependence use specific hypothesis over-emphasising the role of lock-in. With some alterations, however, change can be accounted for (for a model of technological change in the face of network externalities, see Witt 1997). This change knows only suppression of the potential new technology (industry) or a full replacement of the old one. Situations of co-existence between both (e.g. two technological standards sharing the market) cannot be explained.

2.5 Summary and critique

The current chapter has provided an overview of the models relating the development of clusters to the duality of change events and stabilising forces attributed to agglomeration externalities. In all cases, the development of a cluster (i.e. a concentration of firms exceeding that attributable to location-intrinsic factors) is started by some (series of) event(s) enabling a critical mass of local firms. The latter then leads to the emergence of agglomeration externalities stabilising the local industry and driving its growth towards the area's carrying capacity. At some point in time, this process can be disturbed by yet another change event leading to the cluster's exhaustion. The development of a cluster thereby proceeds through a period of stability located between external events. In other words, emergence and exhaustion are attributable to change while endurance is credited to agglomeration externalities. The main difference between the approaches of Sect. 2.1-2.4 then lies with the sources of change as well as the nature and modelling of agglomeration externalities.

The problem with the existing approaches to cluster development in the context of the study pursued here lies in their determinism. To every type of event impacting on an existing cluster, there is only one response. Put differently, if the change benefits the cluster, it grows. Any challenge due to external developments is answered by a reduction of local economic activity. In this setting, given an adverse change event, cluster decline is inevitable. As a result, existing clusters in an industry would have to respond identically, which clearly contradicts the actual evidence (see Sect. 1.1.3). But what are the causes of this determinism? One of its' sources can be attributed to limiting the effects of agglomeration externalities (Morrison and Siegel 1999) to pecuniary, i.e. cost effects. Including positive effects from co-location beyond (transaction) cost savings would 'cushion' a cluster against some of the adverse effects stemming from low cost competitors or changes in trade costs. Second, the view of firms as representative units does not allow for any strategic action on their behalf that might be crucial in deriving adaptation strategies.

The question of how firms or more generally agents in a cluster cope with local bifurcations therefore requires a different modelling approach. The latter would have to be able to analyse the effects of external events in a situation where agents are able to act strategically while being affected by the activities of others due to the existence of agglomeration externalities. Put differently, the model would have to link agent motivations, activities and interdependencies with macro-level outcomes for the cluster (Schelling 1978). In order to advance such a modelling approach, however, the nature and effect of agglomeration externalities on cluster agents first have to be understood in more detail. This is achieved through a review of those concepts that explain the endurance of clusters with the positive effects resulting from agglomeration economies and other benefits to co-location in a more detailed way. The following chapter takes a closer look at this literature that started with the pioneering work of location choice theory and Alfred Marshall in the early 19th century. Building on the insights gained in this part, a model framework for analysing cluster adjustment to change will be introduced in part III.

3 The nature of the beast – On the notion of agglomeration externalities

The theories reviewed in the previous chapter attributed the endurance of clusters to the existence of positive and negative agglomeration externalities, i.e. benefits and downsides to firm co-location. Both were generated by the activities of local firms. As an example, greater numbers of firms in an area increased the availability of labour (through training and in-migration). Agglomeration externalities thus affected locational characteristics (e.g. wages) as well as incumbent and new firm behaviour. Their existence implies that some firm activities impact on the (outcome) of other firms' activities. As a result, agglomeration externalities generate interdependencies between the activities of local agents insofar as the latter's outcome does not only depend on the nature of individual activities but also on the (results of) activities performed by other actors in the area.[25] In the models reviewed so far, this interdependence was limited to cost effects, which will change with the upcoming concepts. Cluster emergence and exhaustion were then attributed to change events in their environment. Events challenging an existing cluster would thus come at a point in its development where the cluster was characterised by a multitude of agglomeration externalities.

If one adopts the aforementioned notions on the effects of agglomeration externalities and their existence at the time of the adverse change, they will also play a role in the study of cluster adaptation. As will be argued in more detail later, adaptation to local bifurcations in clusters proceeds through the activities of local agents. If the latter are interdependent due to agglomeration externalities, the success of individual agent adaptation depends on (the success of) others. This aspect in turn demands a closer investigation of the nature and mechanisms underlying agglomeration externalities in order to understand, which agent activities are interrelated and how this interdependence comes about. Both aspects are provided in the literature explaining the success of clusters as a result of agglomeration externalities. A review of this literature is however problematic due to ambiguity and substantial overlap between concepts presented as distinct schools of thought: "*Few would, even at the outset, feel tempted to accuse the cluster literature at large for being overly concerned with precise definitions or burdened with excessive specifications of the exact nature of the processes involved*" (Maskell and Kebir 2004, p. 3 – see also Martin and Sunley 2003). In order to avoid this issue, the

---

[25] As was already mentioned by Marshall 1920, p. 271: "*[...] if one man starts a new idea, it is taken up by others and combined with suggestions of their own; and thus it becomes the source of further new ideas*".

chapter proceeds as follows. Sect. 3.1 introduces an overview of the effects of co-location in the most general sense. It proceeds by classifying the benefits from co-location into two classes that have not always been distinguished in the existing literature (Moulaert and Sekia 2003). Geographic proximity is argued to offer *first-order/ geographic benefits* in accessing location-intrinsic resources (e.g. raw materials) as well as in cost reductions regarding transportation and inter-organisational exchanges (transactions). In addition, proximity can lead to the emergence of *second-order/ agglomeration benefits* if positive externalities imply an effect of some *activities* of local firms on those of others.[26]

The interesting aspect offered by the cluster literature in contrast to the models presented in the previous chapter is that it includes two types of externalities: *"[Following Scitovski 1954], it is now customary to [distinguish between] two categories: 'technological externalities' and 'pecuniary externalities'. The former deals with the effects of nonmarket interactions that are realized through processes directly affecting the utility of an individual or the production function of a firm. In contrast, pecuniary externalities are by-products of market interactions: They affect firms or consumers and workers only insofar as they are involved in exchanges mediated by the price mechanism. Pecuniary externalities are relevant when markets are imperfectly competitive, for when an agent's decision affects prices, it also affects the well-being of others."* (Fujita and Thisse 2002, p. 8). This aspect has a far-reaching consequence for the study of cluster adaptation to change. The limitation of agglomeration benefits to cost-saving pecuniary externalities was a prime cause of the deterministic behaviour of clusters in the models reviewed in Chap. 2. If firm location choice depends on the profitability of being in an area and if clustering only has an effect on firm production cost in the widest sense, changes in model parameters affecting either of the two aspects will have deterministic consequences. If change increases locational profitability, new firms locate in the cluster and existing ones grow, i.e. the outcome is further agglomeration. If change decreases locational profitability, firms exit the area or the market, i.e. the outcome is a spatial dispersion of enterprises implying cluster decline.

The notion of technological externalities challenges this determinism. If clustering affects local agents beyond mere cost aspects, positive technological effects might compensate for an adverse change in cost-related parameters. For example, if being in a cluster has a strong role for firm innovation, the emergence of low cost competitors does not lead to a straightforward decline of the cluster. Technological externalities thus increase the role of agent activities in cluster development: On the one hand, they may stabilise the existence of the cluster in the face of smaller change events. On the other hand, changes in agent activities affecting technological externalities might result in their loss and speed up decline.

---

[26] The terminology has a direct connection to the theories of path dependence discussed in Sect. 2.4 where the benefits to being in an area were composed of geographic benefits (in absence of other firms) and agglomeration ones (depending on the local firm population). Here, however, the distinction line is slightly different. Geographic benefits arise from lower transaction and transport cost in accessing location-intrinsic resources, whereas agglomeration benefits emerge due to positive externalities (interdependencies) between local firm *activities*.

An understanding of the nature and mechanisms underlying the two types of agglomeration externalities is of prime importance for the study conducted here. Sect. 3.1.1 reviews the positive pecuniary and technological agglomeration externalities ('agglomeration economies') that were introduced by Alfred Marshall. Sect. 3.1.2 highlights the contribution of later research with respect to the nature of and the mechanisms underlying different 'Marshallian' agglomeration economies. In this context, it is found that many activities underlying the emergence of agglomeration economies exhibit a (partial) public good phenomenon. This implies a need for a regulation mechanism to avoid a 'tragedy of the commons'. The latter is usually argued to lie with an emerging local institutional infrastructure (local culture). Sect. 3.2 then introduces the different concepts that have attained prominence in the cluster literature (Industrial Districts, Porterian Clusters, Regional Innovation Systems and Innovative Milieux). The distinction between these concepts is often difficult due to both conceptual overlap and frequent ambiguities regarding the exact definitions of terms. As a consequence, no clear-cut distinction between them will be obtained. However, it is possible to delineate the concepts according to the emphasis placed on different agglomeration economies and the role of the local culture as well as the main firm activities affected by both (Sects. 3.2.1-3.2.4).

What will emerge from the review of both the general mechanisms and the respective concepts proposed in the literature (Sect. 3.3) is that too great an emphasis has been placed on positive agglomeration effects while neglecting the role of negative externalities. Moreover, the literature suffers from stasis in the sense that it almost exclusively focuses on factors making a cluster successful at a specific point in time. "*Many of the available cluster studies in the literature have been focused on more static descriptions of their characteristics at a given point in time, although flavoured with evidence of some of the main features of their history.*" (Dalum et al. 2002, p. 2), while: "*The hallmark of successful [clusters] is change and innovation, not stability and rigidity*" (Staber 1996a, p. 308). In order to introduce a more dynamic perspective into the cluster literature, however, Chaps. 4 and 5 will argue the need for a different modelling approach than those presented in Chap. 2, which is due to the existence of both pecuniary and technological externalities as well as the behavioural impact of the local culture.

3.1 The advantages from co-location

As has been outlined in the introductory section, clusters are understood as non-random spatial concentrations of economic activity. In order to explain their existence, two aspects have to be met. First and foremost, local agents must have come to choose the same area as the site of their operations. Second, their co-location has to present agents with benefits if the spatial concentration of economic activity is to exceed the level attributable to differences in area-intrinsic aspects. As a consequence, the existence of clusters is tied to both an identical location choice of agents and the ensuing advantages from co-location.

The strand of theory investigating the rationales of firm location choice was introduced by the works of Johann Heinrich von Thünen 1826 and Alfred Weber 1909. While the theory acknowledges that locations can come to be chosen due to personal or 'non-rational' motives, the latter's influence is argued to 'cancel out' once the locational dynamics of an entire industry are studied (Ross 1896). Within location choice theory, a site is therefore chosen according to economic reasons, i.e. it comes to host production if *"the selected locality [is] deemed to offer certain advantages in production or marketing over any other equally available point"* (Ross 1896, p. 247). In the traditional (neoclassical) theory, these economic reasons amounted to differentials in production costs, i.e. firms were argued to locate where production (including transport cost) was cheapest.[27] The economic reasons underlying the location choice of firms could then be understood more generally as related to the (net) benefits offered by an area. These depend on how attractive it is to be local, i.e. how much firms benefit from being near area-intrinsic resources or other local agents. Both aspects lead to the notion of *proximity benefits*.

Proximity benefits can broadly be classified into two categories: geographic and agglomeration effects. The first-order, geographic effects are attributable to the advantages resulting from spatial proximity in its own right, i.e. independent from the activities of local agents. They include 'location-intrinsic' factors[28] such as raw materials, climate as well as political incentives like investment subsidies or favourable regulation. Furthermore, being spatially close to resources, other firms, or consumers results in savings on transportation as well as transaction costs. The latter refer to the cost of inter-organisational exchange and encompass costs occurring *"before (search and information cost), during (bargaining and decision costs), and after (policing and enforcement costs)"* a transaction (Maskell 1999, p. 4; Coase 1937). Proximity alone helps lower the first two elements of transaction costs as the parties involved in an exchange can find one another and meet both more easily and frequently when located near each other.

The second-order agglomeration effects then regard the effects emerging from the activities of co-located firms, i.e. they occur when activities by one agent affect others. These effects can be positive and negative, i.e. they encompass the often-cited agglomeration economies (Marshall 1920) as well as diseconomies or costs to clustering (congestion effects). In the existing literature, agglomeration economies feature by far more prominently than congestion effects. As a consequence, they will be discussed in more detail here. However, congestion effects constitute the 'other side of the coin' of agglomeration economies. Following the notion proposed in Chap. 2, they will be viewed as a limitation to the latter and thus act as a growth constraint for clusters.

---

[27] Later developments included other factors into the analysis, e.g. bounded rationality (Simon 1955), multiple goals among individuals in an organisation (Townroe 1972) or interactions with surrounding institutions (Holmes 1999). See also Christaller 1966; Engländer 1924; Lösch 1962 or Palander 1935. For overviews of location theory view Hayter 1997; McCann and Sheppard 2003 or van Dijk and Pellenbarg 1999.

[28] Especially prior to the industrial revolution, these factors explained spatial concentrations of companies (Crafts and Mulatu 2004). Natural advantages continue to be important in industries depending on physical inputs that are hard to transport (e.g. mining and steel).

Regarding the role of geographic and agglomeration effects, it has to be noted that the second-order benefits depend on the presence of a number of local agents conducting interrelated activities, i.e. one cannot postulate that *'agglomerations form because of agglomeration economies'* as has often been done in the literature, since the latter are created by the former (Fujita et al. 1999, p. 4). It is therefore suggested (following Storper 1995), that geographic benefits are responsible for offering the potential of creating an agglomeration of industry actors, while agglomeration effects come to be important in established clusters. Industries exhibiting characteristics offering a significant role for first-order proximity benefits would therefore be more prone to choose the same site of operations, either because location-intrinsic factors favour one area over (most) others or because transaction and transport cost savings make locations with a concentration of other industry agents (Steinle and Schiele 2002) or consumers (Marshall 1920) attractive. A potential for clustering is however not a sufficient condition for cluster emergence unless it is found with a key role for scarce, location-intrinsic factors (e.g. raw materials shaping the spatial distribution of extracting industries). In other industries, historical accident (e.g. regarding the location choice of pioneering firms acting as good examples to start-ups and as a potential source of spin-offs) will have a lot to say about whether and where a cluster emerges.[29] The existence of agglomeration economies then leads to a degree of spatial concentration exceeding the level warranted by location intrinsic or transaction/ transport cost related factors. Put differently, agglomeration economies enable the emergence of clusters as compared to mere agglomerations of economic activity.

3.1.1 It all started with Marshall: Agglomeration economies

The research on the *"permanent advantages [...] to firms located in close juxtaposition to other similar and related firms"* (Malmberg and Maskell 2002, p. 432) was introduced in the early 19th century in Alfred Marshall's book, 'Principles of Economics' (first published in 1890). Observing the economic world around him, *"Marshall quickly came to the conclusion that, at least for certain types of production, there were two efficient manufacturing systems: the established method, based on large, vertically integrated production units, and a second one based on the concentration of many small[er] factories specializing in different phases of the same production process and [operating] in one location or in a cluster of locations"* (Becattini 2002, p. 84). Marshall originally attributed the greater competitiveness of clustered (small) firms over large integrated companies to three different factors (Marshall 1920, pp. 280-284). First, the small size of firms implied better control of employees, easier communication (less people/ hierarchies) and

---

[29] As a result, different industries exhibit distinct clustering potential and dynamics (Hoover 1937; Swann et al. 1998). On evidence of firm location choice and cluster emergence, see among many others Anderson 1992; Arthur 1990; Audretsch and Fritsch 1999; Aydalot 1986; Klepper and Sleeper 2004; Klepper 2002; Lemarie et al. 2001; Sorenson 2003; Storper 1995 or Stuart and Sorenson 2003.

less waste of material (costs felt more directly in small firms). Additionally, their mode of production with independent producers specialising in different stages of the value chain allowed for a greater flexibility regarding product characteristics and output quantities (Goldstein and Gronberg 1984). This was achieved by changing combinations of suppliers and end-producers. Third and most importantly however, it was argued that the spatial concentration of firms leads to the emergence of external economies of scale: If the activities of individual firms add up in a way that benefits the entire local industry (due to efficiency advantages), demand for individual firms will also increase.[30]

A special emphasis in the literature was therefore placed on the causes and effects of agglomeration economies. Usually, two kinds are distinguished: Those positive externalities deriving from proximity to industries and services in general (urbanisation economies; Hoover 1970) and those deriving from the *"geographical agglomeration of related economic activities"* (localisation economies; e.g. Maskell 2001, p. 922). Urbanisation economies in the sense of agglomeration economies across sectors are usually found in areas with a more diverse industrial mix such as cities (Jacobs 1969).[31] Localisation economies tend to underlie the formation of industry-specific agglomerations – i.e. clusters. Although the verdict on the relative importance of either type of agglomeration economies is still pending (Fujita and Thisse 2002, p. 267), this study focuses on the case of clusters and therefore investigates agglomeration economies as localisation effects. The importance of localisation (agglomeration) economies for the cluster then lies with their stabilising effect: *"When an industry has thus chosen a locality for itself, it is likely to stay there long: so great are the advantages which people following the same skilled trade get from near neighbourhood to one another"* (Marshall 1920, p. 271). Put differently, *"agglomeration generates inertia [...], people and firms are there because other people and firms are there too. So people are willing to move out of the agglomeration only if a large shrunk of people are willing to do as well"* (Robert-Nicoud 2004, p. 3). To some extent, agglomeration economies therefore sustain a cluster against smaller external influences such as changes in production, transport or transaction costs as any new area wanting to host the industry would have to offer benefits exceeding the 'matrix' of the benefits generated by agglomeration economies in the existing locale as well as the cost of firm-relocation: *"[t]he power of a locality to hold an industry, [...] greatly exceeds its original power to attract. [Any] new locality must not only excel the old, but must excel it by margin enough to more than offset the resisting power of the matrix"* (Ross 1896, p. 265).

---

[30] This distinguishes external economies from internal ones. Both derive from an increase in production volumes, but in the case of external economies, that increase in production is owed to an overall improvement of local industrial activity whereas internal economies are realised when a company can raise production levels through improvements of in-house resources, organisation or management efficiency (Marshall 1920, p. 277).

[31] Some sources argue that some cross-sectoral urbanisation economies, especially cross-sectoral information spillovers are a very powerful localisation driver as they are even more sensitive to distance than intra-sectoral externalities (Audretsch and Feldman 1999; Autant-Bernard 2003).

The nature of agglomeration economies is usually equalised to the advantages accruing to co-located firms as a result of the 'classical' Marshallian externalities (Marshall 1920, p. 271):[32]

1. *"The mysteries of trade become no mysteries; but are as it were in the air"*;
2. *"the economic use of expensive machinery"*;
3. *"subsidiary trades grow up in the neighbourhood"*; and
4. *"in all but the earliest stages of economic development a localized industry gains a great advantage from the fact that it offers a constant market for skill"*.

The existence of Marshallian agglomeration economies therefore implies that the activities of firms positively influence those of others (e.g. firms' training activities contribute to the availability of skilled personnel). However, the exact mechanisms underlying the different externalities require further elaboration.

3.1.2 Beyond Marshall: New developments in the cluster literature

Subsequent research on clusters has focussed on a clarification of Marshall's descriptions both regarding the meanings of and the mechanisms underlying agglomeration economies. However, only three of the four original factors entered the subsequent analysis as *Marshallian externalities*. They are usually labelled *information spillovers (1), common production factors (3)* and *pooled labour markets (4)*.[33] Furthermore, analysis came to be extended to different possible 'structures' of clusters beyond the 'industrial districts' of small and medium sized firms, i.e. an extension was made to include all spatial concentrations of industry actors that give rise to agglomeration externalities.

With respect to the mechanisms underlying different agglomeration economies, the following notions have been advanced. The emergence of a pooled labour market results from two effects. On the one hand, talented people might be attracted to moving into the cluster due to better employment opportunities. On the other hand, training activities by a greater number of local firms imply that more industry-trained personnel are available locally. In a similar vein, the emergence of specialised suppliers is spurred by a greater market for industry inputs in the

---

[32] Regarding the aforementioned distinction between technological and pecuniary externalities, one can say that the classical Marshallian externalities are partly technological (1) and partly pecuniary (e.g. (4) having a cost (wages) and a quality (training) component).

[33] Empirical studies have found evidence of Marshallian agglomeration economies. Some studies focus on the role of knowledge spillovers for innovative activity (e.g. Anselin et al. 1997; Audretsch and Feldman 1996; Audretsch and Stephan 1996; Baptista 2000; Baranes and Tropeano 2003; Beal and Gimeno 2001; Bottazzi and Peri 2002; Breschi 1998; Feldman 1999; Fritsch and Franke 2004; Keller 2000; Koschatzky 1998; Oerlemans and Meeus 2002; Oerlemans et al. 1998; Orlando 2000; Shefer and Frenkel 1998 and of course Jaffe et al. 1993), although this explanation is disputed (Helsley and Strange 2002). More encompassing tests are provided by Henderson 2003; Le Bas and Miribel 2005; Morrison and Siegel 1999; Rosenthal and Strange 2001, 2003 or Wheeler 2003.

cluster. This ensures a better and cheaper availability of industry-specific inputs. Such a division of labour between local firms allows for the emergence of specialisation and (external) scale effects: As firms focus on one stage of the production process, they can manufacture greater product quantities (scale) in a more efficient way (specialisation) than integrated producers of the same size. If these firm-level effects further add-up in such a way that the cluster as a whole benefits (e.g. due to efficiency advantages), any increase in demand for local products will convey an additional external scale effect to individual firms.

For both kinds of agglomeration economies to materialise, however, a certain degree of understanding between local firms has to be obtained. This is due to the fact that the results of some activities leading to agglomeration externalities have a (partially) common good character. Trained personnel cannot be bound to one employer forever, thus opening up an avenue for individual firms to free ride on others' training investment by hiring trained workers without contributing to the pooled labour market through their own training investment. Individual firms can therefore only be expected to engage in training activities, if they can capitalise on their investment at least for some time. Specialised suppliers in turn will only emerge if they can trust to sell their inputs at a fair price: *"[t]he one who makes the heads of the pins must be certain of the co-operation of the one who makes the points if he does not want to run the risk of producing pin heads in vain.*[34] *The labour operations of all must be in the proper proportion to one another, the workmen must live as near to one another as possible, and their co-operation must be insured"* (List 1909, Book II, Ch. XII, p. 7).[35] A regulation mechanism is hence required to avoid a 'tragedy of the commons' with respect to the activities underlying agglomeration economies, i.e. the exploitation of individual actor effort by free riders in the area has to be limited or stopped. Otherwise, the very activities enabling positive agglomeration externalities would cease. This regulation is obtained by an emerging set of local institutions in the guise of formal and informal 'rules of the game', rewarding compliance and punishing defection. They establish incentives for conducting activities that lead to agglomeration externalities.[36]

It is argued that geographic proximity facilitates the emergence of such a set of institutions (also termed *local culture*) for two reasons. First, actors within a locale

---

[34] *"Der, welcher die Köpfe der Nadeln macht, muß der Arbeit dessen, der die Spitzen macht, gewiß seyn, wenn er nicht Gefahr laufen soll, umsonst Nadelköpfe zu fabricieren."* (List 1842, p. 224).

[35] List 1909 argues that geographic distance hampers this division of labour: *"Let us suppose e.g. that every one of these ten workmen lives in a different country; how often might their co-operation be interrupted by wars, interruptions of transport, commercial crises, etc.; how greatly would the cost of the product be increased, and consequently the advantage of the division of operation diminished; and would not the separation or secession of a single person from the union, throw all the others out of work?"*

[36] This corresponds to the notion in the new institutional economics: Institutions (formal or informal) arise since *"the parties to an exchange wish to economize on transaction costs in a world in which information is costly, some people behave opportunistically, and rationality is bounded. [...] institutions reduce uncertainty by providing dependable and efficient networks for economic exchange."* (Powell and DiMaggio 1991, p. 4; North 1993).

share a set of formal institutions such as legislation as well as some general informal institutions like norms deriving from the respective cultural framework. As a consequence, agents in an area are more likely to judge the same behaviour as cooperative or defective (Glaser et al. 1999; Granovetter 1973; Louch 2000; Sweeney 1995; Varian 1990). Second, the better observability of individual behaviour in proximity facilitates the identification of defective action. This stems from the fact that people can meet more frequently when they are spatially close as travel costs are negligible. Moreover, agents in an area often also interact in beyond the mere business sphere (Burt 2001; Dahl and Pedersen 2003; Lorenzen and Foss 2003). Therefore, location-specific institutions – a local culture – can evolve and are supported more easily in proximity as individuals can observe, judge and react collectively upon different kinds of behaviour (Holländer 1990). They can thus punish defecting agents by excluding them from all future exchanges. Since cluster agents interact in multiple business networks (Cappellin 2003), defection in one entails the risk of losing other valuable ties due to collective punishment. This greatly increases the cost of defection ensuring compliance with the local rules of the game.

However, not all clusters are characterised by such an informal punishment mechanism. In many instances, the threat of using formal institutions, e.g. through legal action being taken substitutes for more informal agreements (Caeldries 1996). A third possibility for arriving at a shared understanding between cluster agents as to what constitutes acceptable behaviour lies with the formation of more durable alliances between some organisations or with an emerging hierarchy of agents in the sense of different propensity of cluster actors to influence the decisions of others (for further detail, see Chaps. 4 and 7). If an agreement on acceptable business practice has been obtained, it further lowers transaction cost between actors: Behaviours that are commonly found unacceptable do not require a contractual treatment, thereby reducing the policing and enforcement part of transaction costs. The lower transaction costs in turn can give rise to a greater degree of division of labour between local industry participants and thereby fosters the specialisation and scale benefits mentioned by Marshall.[37]

Agglomeration economics supported by some form of local culture thus increase local firm's competitiveness relative to that of outsiders. This enables clusters of independent firms to produce at a similar efficiency as large, integrated companies. However, it does not explain why the area would host many independent companies instead of one or a few large, integrated ones. Consequently, the relative competitiveness of either mode of production is of key importance or: What are the advantages of "*N co-localized firms of size S undertaking activities [over] a single firm of size N x S doing the same*" (Maskell 2001, p. 927)? In order for such an advantage to exist, the effects of some agglomeration externalities

---

[37] It has been argued that the division of labour deepens as the agglomeration evolves (Malmberg and Maskell 2002). A two-fold relationship between division of labour and clustering therefore exists. A strong need for a division of labour in an industry can foster agglomeration (Steinle and Schiele 2002). Moreover, agglomeration can allow for a greater division of labour (in all kinds of industries) than the one obtainable over distance.

have to be non-linear, implying that *"the change in a variable is not simply pro-portional to its size [... but] reflects some 'collective' behaviour of some kind which affects individual [elements], so that they react 'faster' or 'slower' than they would [given the same environmental conditions] if they were alone"* (Allen 1997, p. 10). Put differently, the effect of being in a cluster has to improve firm competitiveness as compared to isolation.

The first two agglomeration economies (pooled labour and input markets, firm level scale economies in a stage of the value chain) are often argued to help enterprises in clusters compete with larger, integrated firms by offsetting their size disadvantages regarding resource endowments and scale economies (Becattini 2002; Marshall 1920, Ch. XI). In turn, while the local culture helps reduce transaction costs in exchanges between co-located firms, a large firm conducting all value chain activities internally would not incur these in the first place. Both effects therefore do not provide an explanation of why *"the total economic effect of cur-tailed information and transaction costs as well as of scale advantages [in large firms] are inferior to the locational economies being available when being sepa-rate firms"* (Maskell 2001, p. 927). It would therefore seem that the superiority of clusters as compared to integrated firms arises from non-linearities in the effects of the first Marshallian externality, information spillovers.

The notion of information spillovers has been used to describe a lot of different phenomena, which require careful sorting. It will be argued here that information spillovers consist of two very distinct externalities (Dahl and Pedersen 2003). On the one hand, they refer to the ease with which 'news' about the general business environment and individual actor activity spread within an area due to the prox-imity of actors. As a result, each actor is better informed about what happens in the sector than he would be through individual information acquisition alone. This is similar to the creation of a 'pooled' market for general business-related informa-tion, which could be argued to offset some of the disadvantages of smaller firms, as they are usually less able to free resources and personnel for extensive envi-ronmental monitoring. On the other hand, information spillovers in a cluster can affect more 'core' firm capabilities by offering better opportunities for inter-firm learning. Improvements in inter-firm learning can in turn stem from two different sources: Learning from observation and learning from interaction (Maskell 2001).[38] The first kind of learning refers to the effect of different local actors com-peting in the same line of business under identical local conditions. These agents experiment with different solutions to common problems, which can easily be compared due to very similar production conditions as well as the aforementioned availability of more information about what others in the cluster are doing. This evaluation of the relative strength and weakness of each firm's solution in turn en-ables an imitation of best practice. The second type of learning stems from interac-tion with other firms offering complementary activities (e.g. in R&D co-operations). It benefits from the lower transaction costs attributable to firm co-location and the local culture of the cluster.

---

[38] For evidence of the role of learning from observation in the biotechnology industry see Powell et al. 1996 or Powell 1998.

The non-linearity in the effects of knowledge spillovers is argued to arise in both learning effects between firms in the cluster. The learning from observation of competing firms experimenting with different solutions cannot be replicated in integrated organisations. It originates with differing visions of agents regarding the best solutions to a problem, which would be detrimental when found within the same firm (Loasby 2000; Maskell 2001). The efficiency of learning between collaborating firms is also greater than within an integrated firm. Unlike different departments within one organisation, independent agents have to compete for co-operation partners with other firms conducting the same activities.

Both types of learning in clusters therefore provide a non-linear effect ensuring that in some industries "*the full economies of division of labour can be obtained by the concentration of large numbers of small businesses of a similar kind in the same locality; [... rather than] by production on a large scale*" (Marshall 1920, p. 277). The concepts presented in the following section address this aspect by drawing to differing extent on agglomeration economies and the notion of a local culture in explaining the superior performance of such clusters of firms.

3.2 On Districts, Porterian 'Clusters', Innovation Systems and Milieux

With the rise of large, integrated firms, the Marshallian reasoning on small enterprise clusters disappeared from the centre stage of economic research for a considerable amount of time. It was only during the crisis of the large companies in the early 1960s, that the notion of clusters began to re-emerge in (mainstream) economics (Piore and Sabel 1984). This development was started by research on regions in what came to be termed the '3rd Italy' (Sect. 3.2.1).[39] With the introduction of Porterian 'Clusters' (3.2.2), the literature came to occupy its current central position in economic theory and especially policy. This process was reinforced by the parallel work on the role of space in firm's innovative activity, mirroring the development of the modern economy towards the (in)famous 'knowledge economy'. The growing importance of intangible assets, particularly intellectual capital increases the importance of knowledge externalities, especially for more elaborate activities (Jovanovic 2003, pp. 52-53). The concepts of regional innovation systems (3.2.3) and innovative milieux (3.2.4) address this issue by assessing the role of agglomeration economies and local cultures for firms' innovative activities.

3.2.1 Italian Industrial Districts

The Marshallian argument of external economies in areas with large numbers of co-located (small) firms was rediscovered in Italy in the late 1960s. In the time of economic decline accompanying the crisis of the large firm in many nations, some

---

[39] According to Storper 1995, p.193, the term was coined by Bagnasco 1977.

regions in Northeast-Centre of Italy continued to prosper (see also fig. 3.1.). These areas were composed of small and medium sized enterprises (SMEs), specialising in different stages of the value chain and producing a final product in flexible networks of suppliers and end-producers ('decentralised production').

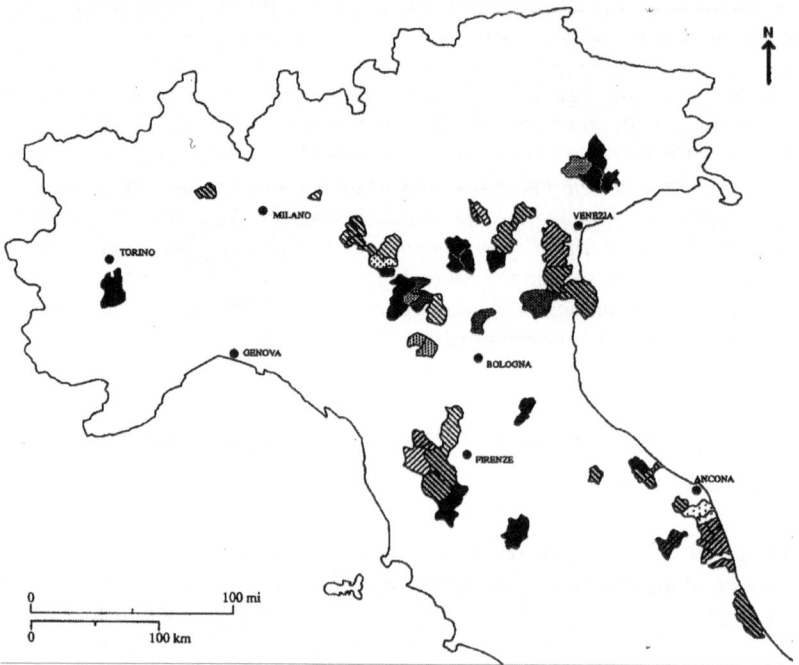

Fig. 3.1. Italian Industrial Districts – Main locations
Source: Sforzi 1990, p. 85

The question of why these SME clusters were able to compete with large integrated firms then became the driving force in the literature. In order to explain the phenomenon theoretically, researchers (e.g. Becattini 1990, 2002; Brusco 1990; Dei Ottati 1994b; Pyke et al. 1990) turned to Marshall's agglomeration economies and developed the soon largely popular notion of 'Italian Industrial Districts' (henceforth IIDs). They are defined as areas with *"geographically concentrated small and medium sized firms targeting their products at the upper market segment where they possess a competitive advantage regarding their flexibility and specialisation. This advantage is obtained through decentralised production in specialist firms with vertical cooperation and horizontal competition. A supportive social environment enables this mode of production and sustains it against economic crisis. The nature and success of a district very largely depends on its his-*

tory which has created the local value system lying at its very core" (Pyke and Sengenberger 1990, p. 2-3).[40]

IIDs derive a competitive advantage in their fashion-driven or high product-customisation markets from the specialisation of independent producers and the flexible recombination of manufacturing networks allowing for a greater flexibility regarding both product characteristics and quantities. The main Marshallian externalities underlying this division of labour are therefore the emergence of specialised suppliers as well as pooled labour markets ensuring a steady and flexible supply of skilled workers. The role of information spillovers and innovation is often argued to be less central in IIDs due to both the small size of companies and the relative maturity of their main industries.[41]

A greater emphasis in the IID literature is placed on the local culture (resulting from strong social networks and extended family ties; Dei Ottati 1994c) supporting the aforementioned mode of decentralised production. For an individual firm to specialise in one step of the value chain, it must be able to trust in being able to sell its' products at a fair price. Consequently, some degree of co-ordination among district firms is required since "*the output of one firm may be the input of another*" (Malmberg and Maskell 2001, p. 11). Moreover, strong demand fluctuations in IID markets imply a need for flexibility in production volumes (van Dijk 1995; Lazerson 1990). This is achieved by horizontal 'outsourcing', i.e. competitors can share orders that are too large for either of them, or by altering the labour force in individual firms. The local culture ensures that a certain degree of reciprocity is maintained in the horizontal outsourcing and that temporarily laid off workers can find employment in other firms or be sustained by family/ community members until the market situation improves (You and Wilkinson 1994).

As a result, the concept of IIDs was first to emphasise the role of a local institutional framework in sustaining a mode of production leading to agglomeration economies (Harrisson 1992): While Marshall did mention that a supportive environment is needed to start industrial activity in a given location ("*there is perhaps no part of the old world in which there might not long ago have flourished many beautiful and highly skilled industries, if their growth had been favoured by the character of the people, and by their social and political institutions*" Marshall 1920, p. 270) it was not a factor that contributed to agglomeration externalities in his 'districts'.

Subsequent work extended the very specific case of IIDs to other countries as well as to districts that were not exclusively composed of SMEs (Markusen 1996). Alongside their Italian variant, industrial districts can consist of a number of (SME) suppliers organised around a large, firm (hub and spoke industrial district)

---

[40] For empirical studies, see Amin 2000; Amin and Thrift 1992; Becattini 1994; Brusco 1982, Brusco and Righi 1989, Bull et al. 1991, Capecchi 1990, Dei Ottati 1994a; Goodman et al. 1989, Paniccia 1998, Rabellotti 1997, Staber 1996a, You and Wilkinson 1994.

[41] R&D effort is mostly limited to adaptation and improvement of machinery and/ or production processes (Asheim 1996, Lazonick 1993).

or a research institution (state-anchored industrial district).[42] Finally, as was high-lighted in Sect. 1.1.3, the recent problems of IIDs in coping with different external changes has led to a growing number of case-studies on district adaptation (e.g. Amin 1994; Bellandi 1996; Bianchi 1998; Dei Ottati 1996; Paniccia 2002 or Russo 1985).

3.2.2 (Porterian) Clusters

The probably most famous contribution in the cluster literature is Michael E. Porter's book "The competitive advantage of nations" (first published in 1990). Starting from the observation that an industry's leading firms are often found in the same country or even area, Porter identifies four key factors in a nation (region) that enhance firm productivity: Factor conditions, local demand, inter-firm rivalry as well as related and supporting industries. Taken together, they constitute an interrelated 'diamond' (Fig. 3.2.) that can increase the competitiveness of firms in the area.

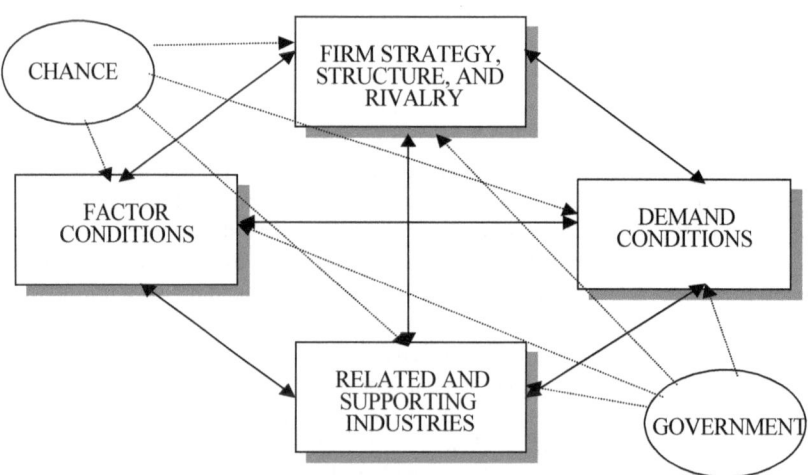

Fig. 3.2. Porterian Clusters – the constituent elements
Source: Porter 1998, p. 127

Originally intended to explain competitiveness at the national level, Porterian Clusters were soon used to describe the existence of regional concentrations of economic activity (Enright 2003). The argument behind this was that the elements of the Porter 'diamond' could be expected to work even more strongly in prox-

---

[42] Furthermore, spatial concentrations of branch plants belonging to outside multinationals are included in the analysis as constituents of satellite platform districts. However, the notion of satellite platforms departs with the original district idea of a key role for localised interactions and agglomeration economies. Instead, the key driver to firm activities and thereby the dynamics of the 'platform' is the mother corporation located outside the area.

imity. The concept then implicitly draws on Marshallian agglomeration external-ities to explain the development of the cluster: The notion of pooled labour mar-kets and specialised supplier emergence in spatially concentrated industries is im-plicit in the assumption that firm clustering influences factor conditions as *"few factors of production are truly inherited [...]. Most must be developed over time through investment"* (Porter 1998, p. 77). The greater the number of local firms, the better the development of industry-specific production factors. Knowledge spillovers in the guise of learning from observation and collaboration are also prominent in Porterian Clusters.

The focus in the concept is however on the contribution of these different ex-ternalities to firm competitiveness. Better factor conditions and internationally competitive suppliers provide firms with access to world-class inputs, while local rivalry pushes them towards increased productivity and innovativeness. Moreover, the agents contributing to competitive advantage extend beyond the Marshallian understanding: The definition of the cluster is broadened to encompass not only firms and local agents such as service providers, but also firms in related indus-tries, industry-relevant institutions like universities or trade associations as well as customers. By the inclusion of related industries and institutions, Porter introduces a notion of both localisation and urbanisation economies as supporting forces in 'clusters'. In addition, customers (home demand) play a special role for competi-tiveness as they generate additional pressure on firms to improve their products.[43] All four elements of the diamond are self-reinforcing. As an example, better factor conditions improve the competitiveness of firms in the industry. As the sector grows, training activity (e.g. in universities) could become more targeted, thereby improving factor conditions, etc.

The local culture in the context of Porterian Clusters is not only about the ques-tion of trust and support to a division of labour as it was the case for the IIDs. In-stead, it encompasses all (voluntary or formal) social constructs supporting the emerging pattern of vertical and horizontal 'co-opetition' (simultaneous co-operation and competition) and especially the learning between local firms.[44] As any cluster evolves against a specific industrial, national and regional background, its' structure and dynamics are difficult to replicate elsewhere, thus binding firms interested in capitalising on the benefits of the cluster to the latter's host area.

3.2.3 Regional Innovation Systems

The role of proximity for inter-firm learning was further emphasised with the in-troduction of Innovation Systems. The concept originates in the effort to explain *"the relations between technological change, the economy and wider society"*

---

[43] Firms underlie a 'home-market bias' in the sense that they react most strongly to changes in demand in their home market. This also holds for multinational enterprises.

[44] Again, there are numerous case studies in the Porterian framework. These are either targeted at the regional or the industry level (e.g. Audretsch and Cooke 2001, Braunerhjelm et al. 2000, Bresnahan et al. 2001, Orsenigo 2001, Swann et al. 1998; Wolter 2004).

(McKelvey 1991, p. 117) and has four different units of analysis: The technology, the sector as well as the nation state and the region. The key idea behind the innovation system approach is a novel conceptualisation of the research and development (R&D) process underlying successful innovation. Traditionally, R&D was viewed as linear: Starting from basic towards more applied science, an innovation would eventually be produced enabling new marketable products or technologies. This process (or at least substantial parts of it) was conducted within individual firms. The innovation system literature challenges this linear view by arguing that R&D and innovation stem from a creative combination of generic know-how and specific competencies. Due to an accelerating pace, growing cost of innovation as well as increasing product complexity, it is now almost impossible to have all the necessary competencies within any individual firm. Additionally, the strong uncertainties about the outcome of any R&D activity provide an incentive for risk sharing and uncertainty reduction by networking with other inventors (Dosi et al. 1988). As a consequence, innovation is increasingly understood as the result of a collective, complex and interactive process (Kline and Rosenberg 1986) carried out in cyclical interaction between different knowledge providers (Edquist 1997; Ratti et al. 1997).

The introduction of this cyclical perspective on R&D marks a shift in the research on clusters. Rather than discussing the effects of proximity on production, emphasis is directed towards the role of space in innovation. If innovation is the result of an interactive process, locational factors affecting the efficiency of its organisation become important. In the context of the innovation system literature, these factors are argued to lie with the formal (legislation, etc.) and informal institutions affecting the co-ordination of economic activity. Put differently, all factors influencing the organisation of innovation networks among knowledge providers are placed at the centre of analysis. The 'system' of innovation then consists of the different knowledge providers (i.e. firms, research/training units and policy institutions; Malerba 1999) as well as their interactions. Both are shaped by industry characteristics determining the main innovation drivers (Nelson and Rosenberg 1993, Lundvall 1995) as well as the specifics of the formal and informal institutional framework in the host area, which govern inter-organisational exchange.

The argument for a territorial focus (national or regional) in innovation systems stems from the consideration that interaction between different knowledge providers will be conducted more easily at a confined geographic level (Cooke et al. 1997; Mothe and Paquet 1997; Urnaga et al. 1998), reflecting both first-order (lower transaction costs) and second-order proximity benefits. The latter regard the institutional framework within a nation or area: If people share aspects like a common history, culture, value system, and language, interaction between them is expected to be easier than with distant, 'foreign' parties (Glaser et al. 1999 or Granovetter 1973, p. 1362). Additionally, the respective nation states (and to some extent maybe also regions) have a role to play in the formal institutions underlying interaction (e.g. legislation on co-operative R&D ventures).

Adopting the perspective of regional innovation systems (RIS) is justified by the fact that regions can be hosts to their own subsets of knowledge providers and that they furthermore differ strongly according to the informal institutions in their respective local cultures (Braczyk et al. 1998). As a consequence, RIS can display different configurations of innovative networks both regarding 'nodes' in terms of participating agents as well as the exact nature of 'links' (interactions) between them. The concept of RIS thus touches upon the idea of Marshallian agglomeration economies in knowledge transfer but focuses to a greater extent on the role of local organisations and geographically confined formal and informal institutions for the efficiency of innovation networks. However, the underlying notion of a role for space in innovation is not uncontested (for a critique see Moulaert and Sekia 2003).

3.2.4 Innovative Milieux

The focus on the formal and informal institutions underlying successful innovation networks is even more pronounced in the concept of innovative milieus. Sharing the understanding of R&D as an interactive process, the main question is on how far the informal institutions in an area's local culture influence the efficiency of innovation linkages between local actors as well as any ties with to non-local agents.[45] An innovative milieu is defined as: *"the set, or the complex network of mainly informal social relationships on a limited geographical area, often determining a specific external 'image' and a specific internal 'representation' and sense of belonging, which enhance the local innovative capability through synergetic and collective learning processes"* (Camagni 1991, p. 3). This definition pinpoints at three characteristics (Courlet and Soulage 1995). First, the milieu exhibits specific behavioural patterns through which independent actors create, process and exchange knowledge. Second, these processes occur both within and between agents (Quevit 1991). Third, there is a strong role for learning from cooperation for local agents.

The institutions shaping the milieu as well as the geographic structure of innovation networks are then argued to contribute to the generation and use of knowledge in the area as a whole (Ratti et al. 1997, pp. 4-5). Knowledge in the broadest definition here is *"the ability to master the production process"* (Maillat, 1998, p. 117). This knowledge in turn constitutes a key resource in adapting to changes in the business environment such as new developments in technology or markets. Thus, the milieu as a determinant of knowledge creation and development can be regarded as one force steering the development of the local industry. In a sense, the innovative milieu literature thereby pays reference to Marshall's notion of knowledge spillovers and the underlying informal institutions (local culture) supporting knowledge exchange, but focuses on the role of both factors for innovation rather than production.

---

[45] The innovative milieu is the only concept within Marshallian cluster theory that pays explicit reference to the role of non-local agents for the dynamics of the cluster. Linkages to the outside world are crucial for obtaining complementary knowledge assets as well as impetus for innovation (Camagni, 1991, p. 5; Quevit 1991).

Critics of the concept state that it does not specify which processes and mechanisms promote innovation (Storper 1995, p. 203). Hence, it cannot explain, why certain regions are more successful innovators than others (Asheim, 1996, p. 393). In addition to that, the innovative milieu aspect is very strongly focused on the process of knowledge transfer between different market participants. In itself, this aspect can however only represent one pillar of the success of regional production systems. Empirically, a connection between the milieu and the innovativeness of firms can only be found for some industries (Shefer and Frenkel 1998).

3.3 Summary and critique

As was laid out in the introductory section, the emergence of a cluster is tied an agglomeration of agents in its host area. The latter can result from historic accident alone but it is usually enforced by some clustering potential found with industry characteristics implying proximity benefits. These can lie with transaction cost reductions in accessing local resources and/ or inter-organisational exchanges. With a critical number of local agents, second-order (agglomeration) benefits in the guise of agglomeration economies and the local culture emerge. Agglomeration economies encompass positive effects from firm co-location on local labour and input markets as well as on the availability of information or inter-firm learning. The local culture consisting of a set of (mainly informal) institutions in turn governs local business practice and thus enables and facilitates the inter-firm exchanges underlying agglomeration economies. With the inclusion of second-order agglomeration benefits, the competitiveness of clusters is no longer based only on productive but also on 'organisational efficiency', i.e. how efficiently inter-firm exchanges work. The analysis has thus progressed "*from a (static) cost geography to a (dynamic) organisational geography*" (Ratti et al. 1997, p. 4). Both effects (agglomeration economies and the local culture) arise at the level of the cluster and cannot be attributed to individual actors inside it. Furthermore, they imply interdependence between agent activities in the sense that agglomeration economies determine how one agent benefits from the activities of others and the local culture favours some kinds of business practice while ruling out other behavioural alternatives as not acceptable.

What is lacking from the existing literature is a consideration of negative externalities ('congestion costs') from agglomeration. For instance, the emergence of new firms in an area can increase infrastructure cost for all agents. Moreover, provided a limit on the stock of available inputs exists, congestion effects may also arise with respect to the cost of production factors and other industry inputs. A third aspect relates to an over-use of natural resources, e.g. excessive pollution of air or water. Following the notion advanced in Chap. 2, congestion costs can therefore be viewed as a growth constraint for clusters, i.e. the trade-off between agglomeration economies and congestion cost limits the number of agents an area can host.

A second and more important problem with the clustering literature in the context of this study regards its lack of dynamics in cluster development. First, through their focus on second-order agglomeration benefits, the concepts presented so far fail to account for how the critical mass of agents necessary to enable the former can come about. Second, there is very limited evidence on the dynamics of existing clusters over time, i.e. the question of how and when the benefits of an area can turn into liabilities is not included in the analysis (see also Sect. 1.1.3).

Summing up the arguments of Chaps. 2 and 3, one could argue that clusters emerge due to a combination of first order proximity benefits and historical accident. Their endurance is owed to the existence of second-order ones. The result of the dynamics underlying cluster emergence and endurance is a cluster with a variety of local actors, as well as a mode of operation, i.e. how economic activities are conducted in the area. Both aspects together form the current cluster configuration which is by no means static in the sense that no micro-dynamics are possible. Examples of the latter include reconfiguration of production networks in response to demand fluctuations as well as the generation and dispersion of new ideas and products. The cluster's stability implies that it has to exhibit sufficient competitiveness to survive within a specific environment (characterised by demand, non-local competitors, the technological domain and of course the host nation). This competitiveness of clusters varies strongly: Many real-world examples exist far from the limelight of the famous, world leading clusters like the Silicon Valley.[46]

Local bifurcations to the cluster's development then stem from changes in their environment leading to a significant reduction of cluster competitiveness. This change requires adaptation, i.e. a change in cluster configuration that goes well beyond the usual micro-dynamics mentioned before. However, when taking the full range of agglomeration externalities proposed in this chapter into account, cluster adaptation becomes a field of study in its own right. With the broader perspective on the effects of agglomeration externalities adopted in the cluster literature, the determinism found in the previous chapter is no longer sustainable. First, in contrast to the models presented in Chap. 2, the Marshallian cluster literature includes technological agglomeration economies, which lead to non-linear effects in the relationship between co-location and inter-firm learning. Non-linearities then imply that small changes in their underlying causes (i.e. agent activities) can have over-proportional effects at the cluster level. Depending on whether agents change the activities linked by non-linear interdependencies or not, different paths of adaptation with differing outcomes become possible. Second, the notion of a local culture means that not all behavioural options might be available for cluster agents in adaptation. Depending to the extent of behavioural restrictions exerted by the local institutional endowment, different adaptation measures are available, resulting again in differing adaptation paths.

---

[46] The environment of clusters in the understanding advanced here includes all factors that are outside the control of local agents but that still influence their competitiveness by impacting on how production can be carried out in the area (technologies and legislation) as well as the market success of products (demand and competition).

As a consequence, another approach is needed to model and study the issue of cluster adaptation to local bifurcations. The former would have to be able to account for agent activities, (non-linear) interdependencies and the possibility of behavioural constraints. Moreover, a certain 'measure' of success of the adaptive moves by agents is needed. Part III introduces this novel approach by linking the characteristics of ideal-typical clusters derived from Chaps. 2 and 3 to those of complex adaptive systems. It thus looks into the promise offered by new tools from complexity theory with respect to a formal description of ideal-typical clusters and the study of adaptation in them by understanding clusters as *complex, adaptive systems*.

Part III Towards a complexity perspective on clusters

> *One of the most highly developed skills in contemporary Western civilization is dissection: the split-up of problems into their smallest possible components. We are good at it. So good, we often forget to put the pieces back together again.*

Alvin Toffler 1984, p. xi

The following chapters address the question of how the issue under study here, the adaptation of agents in clusters to changes in their environment could be analysed from a theoretic perspective. It does to by linking the insights on the nature of clusters outlined in Chap. 3 with the notion that their existence cannot be taken for granted due to the dynamics in the spatial distribution of industries (Chap. 2). Accepting the idea of clusters as composed of interdependent agents without a central decision-making authority however implies that adaptation to change is less than straightforward (Chap. 4). Thanks to the interdependence between the (success of) individual activities and the behavioural restraints exerted by the local culture, agent adaptation does not equal cluster adaptation. Maintaining the notion that adaptation in clusters proceeds in a decentralised fashion through the adaptive moves of agents, the question to be asked then is whether there are any cluster-level factors that can steer individual activities towards 'good' collective outcomes.

Outlining the empirical evidence of real world clusters that have survived in a changing environment shows interesting regularities with respect to what could constitute such versions of 'dynamic capabilities' in clusters. In existing case studies, three factors have repeatedly been singled out as relevant for the success of adaptation:

1. internationalisation of cluster agents,
2. the degree of division of labour between them and
3. the way agent activities are co-ordinated in the cluster.

In a bid to overcome the limitations of case study evidence, chapter 5 then highlights how new insights from complexity theory and especially the N/K model could be used to account for the nature of ideal-typical clusters and to investigate the role of two of these factors for cluster adaptation. The influence of the extent of division of labour (2) and the mode of co-ordination (3) on cluster adaptation will then be analysed through the building of a general theoretic model and simulations (part IV).

4 Clusters, change and adaptation: Sticky places in slippery space?[47]

The different concepts reviewed in Chap. 3 have shown that there is a multitude of forms of and perspectives on clusters. However, a smallest common denominator between them can be derived which will act as the working hypothesis on the nature of clusters for the remainder of the analysis. In all approaches, clusters are viewed as non-random spatial concentrations of economic activity that exist due to the effects of agglomeration externalities. Their emergence is usually tied to a combination of historic accident and industry-specific factors offering a significant importance to first-order proximity benefits. Both aspects can then lead to an initial spatial concentration of public and private organisations: Outside agents can locate into the emerging cluster, existing ones can start engaging in activities relevant to the industry (e.g. universities offering tailored training programmes) and new organisations can be created through spinout or start-up activity.

The parallel conduct of competing and complementary activities by co-located agents then gives rise to second-order proximity benefits in the guise of agglomeration externalities. Agent activities and interactions produce agglomeration economies which in turn affect the outcomes of agent activities in areas as diverse as research, human resources or marketing (among many others).[48] Alongside this, an increasing number of local agents can lead to congestion effects limiting the growth of the cluster. Both types of agglomeration externalities therefore imply an interdependence between the outcomes of local agents' activities (Storper 1995). Agglomeration economies determine how one agent benefits from the activities of others while negative externalities outline in how far the actions by one agent reduce the well-being of others. For instance, if a firm wishes to capitalise on knowledge spillovers from another firm in its R&D activity, the success of the first agent's research depends on his own effort, the research undertaken by the second firm as well as the effectiveness of the transmission channel between both.

It was found that the activities underlying agglomeration economies exhibit a (partial) common good character. As a result, some form of understanding between local agents has to be obtained to limit free-riding and avoid an exploitation of individual effort that would otherwise curtail the very emergence of agglomeration economies (Enright 1995). In the existing literature, this 'understanding' was predominantly argued to emerge with an informal institutional infrastructure in the

---

[47] Quoting Markusen 1996 (paper title).

[48] The relative emphasis placed on the different Marshallian agglomeration economies and the areas of their effects however differs between the concepts investigated.

area. Combined with the easier control and a threat of collective punishment by other local agents, this informal institutional infrastructure was sufficient to sustain 'voluntary' behavioural restraints of an individual actor (Holländer 1990; Kandel and Lazear 1992). However, an understanding between local agents can also be ensured by a credible threat of using formal institutions such as legislation. Dropping the idea of an equal distribution of power between cluster organisation furthermore gives rise to a third form of support for the mutual understanding that is all so key to sustaining cluster dynamics: Alliance or hierarchy. The former corresponds to a scenario where some actors form more closely tied business groups based on long-term (contractual) linkages with some other cluster agents. The latter regards differences in agent propensity to influence or even dictate others' decisions. One or several agents occupying a dominant position in the cluster can ensure an understanding of acceptable practices in the area by using their influence to exclude certain behaviours and favour others.

At the same time, the theories reviewed in Chap. 2 already indicated that clusters are subject to dynamic developments as the spatial distribution of industries underlying their existence may change over time. Such changes were introduced by events external to the cluster. In the existing models, they are accounted for by alterations of model parameters. Due to the limitation of agglomeration economies to pecuniary effects, this produced a change in the costs and benefits to being in any area and thereby resulted in a deterministic development towards greater spatial concentration of the industry (cluster growth) or dispersal (cluster decline). Agents in the cluster had no means of changing this outcome by adaptation. It was argued in Chap. 3 that the inclusion of technological externalities would challenge this result. Having cluster agents linked by both pecuniary and technological externalities implies that their activities are interdependent beyond mere cost considerations. As a result, adaptation in clusters becomes a more complex phenomenon exhibiting non-deterministic outcomes. These are a consequence of the nature of clusters where interdependent agents adapt to external changes while both relying on others' activities and facing behavioural restraints exerted by the shared understanding of acceptable business practice. This conceptualisation requires a new set of methodological tools able to account for the cluster characteristics found with the existing empirical and theoretical literature. To arrive at such a modelling proposition, the following section reviews the nature of clusters and the resulting issues with respect to their adaptation to change in order to arrive at a set of requirements that the model to be introduced in Chap. 5 will have to meet.

4.1 The nature of clusters: Agents, interdependence and co-ordination

An ideal-typical cluster as described in Fig. 4.1. is delimited by relational and territorial boundaries, i.e. it consists of all individuals and organisations in an area that conduct activities within the industry and thus contribute to agglomeration externalities. For the remainder of the study, only the level of organisations (hence-

forth actors or agents) will be considered. In this definition, a multinational's subsidiary would be considered as a part of the cluster provided it interacted with
other local firms. The parent firm, however, belongs to the cluster's environment.
While it could influence the cluster's development through a transfer of assets
(e.g. technologies or management practices) to its subsidiary, it will not be considered as a local agent. Similarly, local political authorities only constitute agents in
the cluster if they conduct activities that are relevant to the local industry (e.g.
provision of research on international markets). *A cluster can therefore be understood as composed of a set of interconnected (public and/ or private) organisations in an industry that jointly produce a (set of) marketable product(s) or ser-
vice(s).*[49] Depending on their position in the local industry's value chain, agents
offer competing (horizontal dimension) or complementary (vertical dimension) activities. In most instances, a cluster will exhibit multiple actors in both dimensions, i.e. there are different groups of agents in the vertical dimension of the
value chain while each group is made up by one or several actors.[50]

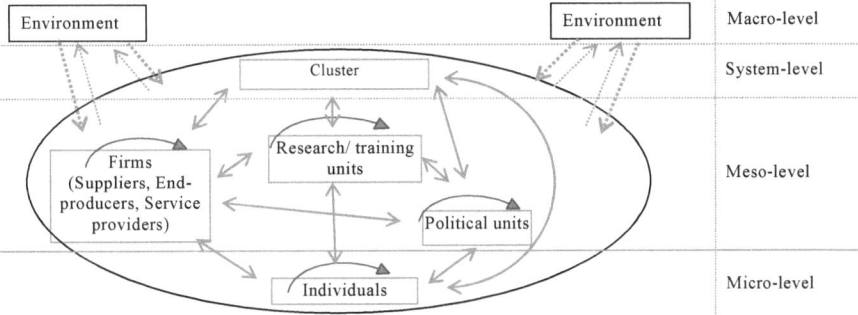

Fig. 4.1. The nature of ideal-typical clusters[51]

With respect to their constituent agents, clusters can differ according to two aspects: The degree of (vertical) division of labour as well as the number, size and
public/ private 'nature' of their agents. In other words, one can find clusters with
and without public sector organisations as well as instances of small or large firm
clusters (Markusen 1996). To some extent, division of labour and cluster composi-
tion are interrelated. The division of labour in a cluster indicates how value chain
activities are grouped within individual agents. In the case of a limited division of
labour, the cluster would be characterised by the presence of fewer integrated

---

[49] See also: *"Der Wirtschaftskreis ist die Verbindungslinie der peripheren Punkte aller
Standortsfiguren, deren Zusammenwirken in der Produktion ein autarkes Wirtschaftsgan-
zes, eine Wirtschaftsverbindung ergibt"* (Ritschl 1927, p. 815).

[50] The competitive stability of this setup with respect to pricing and profitability is found
by Baldwin et al. 2003 or Combes 1997. In a more general context, such cluster structures
are described by Soubeyran and Weber 2002 or Storper and Scott 1988

[51] Please note that in real world cases, not all the agents described here need to be present locally. For a cluster to emerge, a spatial concentration of firms suffices. Put differently, clusters can be composed of firms with or without policy or research/ training units.

agents conducting a substantial share of the value chain in-house. Such actors would in turn tend to be larger than those conducting only a limited range of value chain activities internally. The second type of actors would then be more prominent in clusters with a strong division of labour. One could therefore say that the greater the division of labour (on average), the smaller the range of activities in the value chain performed by individual actors, which would be expected to result in a smaller agent size.

The joint conduct of competing and complementary activities by different cluster agents alongside different (un)planned exchanges between them (represented by the arrows in Fig. 4.1.) then leads to the emergence of agglomeration externalities.[52] These externalities are by no means limited to direct effects in the sense of '*if firm A does x, this implies an effect of y on firm B*'. Instead, they can also result from indirect feedbacks. For instance, pollution effects as one possible source of congestion costs are transferred through 'intermediaries' such as soil or water where the polluting activities of agent A only indirectly reach agent B. Another example regards indirect knowledge spillovers where knowledge leaks unintentionally from one firm to another due to informal conversations among employees. On the positive side, agglomeration economies are usually subject to an indirect (time-lagged) transfer as well. For instance, training activities by the firms in an area increasing the availability of skilled labour can later assist the recruitment of all firms, which increases performance.

Their indirect and sometimes lagged effect furthermore implies that agglomeration externalities cannot be viewed as an intentional goal underlying agent strategies. Just like "*people do not marry to reproduce the nuclear family or work to sustain the capitalist economy*" (Bhaskar 1998, p. 35), cluster firms engage in the pursuit of individual goals, choose their strategies accordingly and then create agglomeration externalities as a by-product of their actions. Nonetheless, the interdependence and interplay of activities conducted by local actors inexorably has repercussions on the dynamics of the cluster as a whole, especially thanks to the partial non-linearity in the effects of some externalities (see also Sect. 3.1.2). This non-linearity ensures that for some industries, clusters of several independent firms perform better than an integrated organisation of the same size. However, non-linear interdependencies between agent activities also imply that changes in individual strategy can have unexpected aggregate effects at the cluster level. This issue will constitute one of the key issues in cluster adaptation (Sect. 4.2).

In a bid to sustain the activities generating agglomeration economies against exploitation by free-riding agents, some extent of co-ordination between the individual cluster actors has to be obtained since the cluster lacks the central decision-making authority found in individual organisations such as firms. "*[T]he one who makes the heads of the pins must be certain of the co-operation of the one who makes the points*" (List 1909, Book II, Ch. XII, p. 7) or any firm investing in research or personnel training has to be assured that its knowledge or employees will not be immediately appropriated and used against it by others in the cluster.

---

[52] For an overview of the (un)planned interactions underlying agglomeration economies, see Cappellin 2003.

As a result, clusters are in need of a *co-ordination mechanism* restricting undesirable individual behaviour through a shared understanding of acceptable business practice. Agent adherence to this shared understanding can be supported through collective observation and punishment mechanisms or by authority relationships between cluster agents (alliance, leader firm structures).

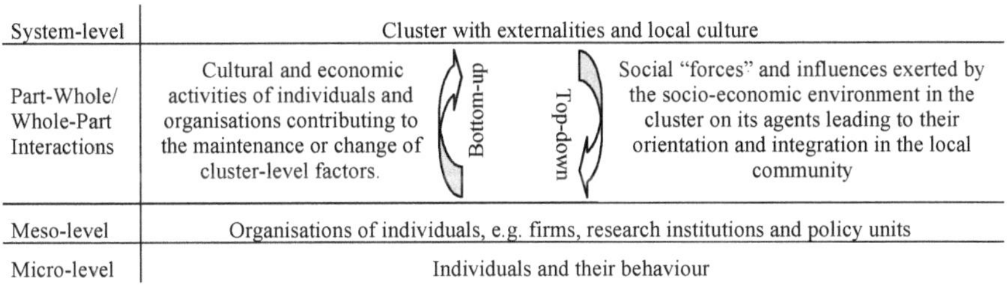

System-level	Cluster with externalities and local culture	
Part-Whole/ Whole-Part Interactions	Cultural and economic activities of individuals and organisations contributing to the maintenance or change of cluster-level factors.	Social "forces" and influences exerted by the socio-economic environment in the cluster on its agents leading to their orientation and integration in the local community
Meso-level	Organisations of individuals, e.g. firms, research institutions and policy units	
Micro-level	Individuals and their behaviour	

Fig. 4.2. Cyclical interaction of micro-, meso- and macro-level in clusters
Source: adapted from Weidlich 2000, p. 13 (with permission from Taylor & Francis)

Both aspects, agglomeration externalities and the co-ordination mechanism sustaining them arise from the activities of all agents in the cluster, i.e. they constitute properties that are not attributable to individual organisations but only to the cluster as a whole[53] (Staber 1996b, p. 152). A cluster characterised by these factors therefore comes to be more than the sum of its parts (agents and agent activities) and constitutes a level of analysis in its own right. The emergence of cluster-level factors in turn gives rise to a cyclical interdependence between its different levels (Fig. 4.2.). While agents and agent activities first give rise to and may later alter agglomeration externalities and the mode of co-ordination between cluster firms, cluster-level factors also influence individual behaviour. Agglomeration externalities determine in how far agents benefit/ suffer from the activities of others while the mode of co-ordination offers "*orientation [...] by extending and also restricting the scope of available information, thought or action*" (Weidlich 2000, p. 12).

To sum up, the development of the cluster leads to the emergence of an architecture characterised by cluster agents, their interdependence from agglomeration externalities and the mode of co-ordination. This architecture is shaped by both the emergence and change of local actors as well as by their activities. Within the cluster's architecture, multiple configurations regarding the presence and activities of local agents are possible. Individual effort alongside the positive and negative effects of cluster-level properties then determine the competitiveness of all cluster agents and the cluster as a whole. Put differently, individual activities and architectural effects shape the competitiveness of any cluster configuration.

---

[53] In psychology, this reasoning is used in cognitive theory: 'Ehrenfels-qualities' or 'whole-qualities' reside in the system as a whole, not in its parts. They get lost when concentrating the observation on the parts. It is argued that system properties crucially depend on the parts constituting the system as well as the relationships between them (Schweizer 1997). A simple example of emergent properties is that of a straight line composed of dots: Straightness is a quality of the system, not of the dots (Kubon-Gilke 1997, p. 349-350).

With the notion of competitiveness, it becomes apparent that clusters are not stand-alone entities but take part in an environment including their host region or nation, product markets, technologies and the likes. Their embeddedness implies another reciprocal influence between individual activities, cluster effects and the environment. In total, the influence of the environment on the cluster is argued to be stronger than the other way round.[54] On the one hand, the success of a cluster depends on the interplay of the results of its current configuration with the environment (e.g. market acceptance of products). On the other hand, the environment influences a cluster through a multitude of channels beyond the market mechanism (e.g. policy intervention, technological developments). In a way, the environmental conditions will therefore influence the emerging cluster architecture and its best possible configuration. Events altering these conditions will challenge the cluster, implying that adaptation (changes in its configuration) has to occur. Adjustment to change can however be constrained by the architectural aspects of the respective cluster that may be hard to change in the short term.

Some properties of ideal-typical clusters (multi-level structure, partly non-linear interactions between agents, emergent cluster-level properties and openness to the environment) coincide with characteristics found in complex adaptive systems. This offers a potential avenue for a better understanding of the behaviour of clusters in the face of change events: If clusters can be described as complex systems controlled by a set of parameters, then their influence on cluster behaviour following change could be analysed. At the end, this offers a way to understand, which factors matter how for cluster adaptation. Before venturing into the analysis of clusters from a complexity perspective proposed here (Chap. 5), however, two additional issues have to be investigated. The first regards existing insight on what parameters might impact on cluster adaptation to change events (4.2) and the second consists of a description of model requirements (4.3) to ensure that an appropriate method within the field of complexity theory is selected.

4.2 Adaptation: Decentralised problem solving by interdependent agents

The previous section has highlighted that clusters exhibit properties that are not attributable to individual agents within them, but only to the collective. Moreover, clusters develop a specific *architecture* characterised by the presence of different types of agents, whose competitiveness depends on both their own behaviour as well as that of others. The range of agent activities is furthermore constrained by the existence of a mutual understanding on acceptable business behaviour acting as a form of co-ordination between actors. This *mode of co-ordination* can be sustained by different mechanisms (collective observation and punishment, threat of using formal institutions, alliance or hierarchy constellations). The combination of cluster architecture (structure, mode of co-ordination) and individual activities

---

[54] The thicker arrows leading from the outside to the cluster (Fig. 4.1.) illustrate this.

then yields a specific cluster configuration which exhibits a certain competitiveness within its environment. If this configuration is challenged by an external event, the adaptation of individual agents will be constrained by architectural factors regarding the range of possible adaptive moves as well as their success (influenced by the effectiveness of individual measures as well as the interdependencies between actor activities). As a consequence, the architecture of the cluster influences the way adaptation is carried out and thereby impacts on the development trajectory following a change event.

This section aims at providing answers to the question of what the success factors for adaptation in clusters are. This involves addressing three related aspects. The first regards an identification of the drivers of change (who adapts) as well as the different architectural contingencies they face (4.2.1). The second aspect to be clarified lies with the timeframe and level of the analysis to be pursued in this study (4.2.3). Third, a review of the empirical literature will highlight, which factors have been found to matter for adaptation in real-world clusters (4.2.2). These insights will be used in the theoretic modelling exercise to determine general causalities that are currently unobservable from case-study evidence. In order to reach its goals, the model will however have to meet a number of requirements that are described in the following section (4.3).

4.2.1 Individual activities or cluster properties: Drivers of adaptation

Studying the success factors for cluster adaptation to external events first involves a clarification of agency, i.e. who adapts? Is it the cluster or does adaptation lie with individual organisations? This question is about who has both the incentive and means to drive adaptation in clusters. Since adaptation is about re-establishing competitiveness, the answer is relatively straightforward: Agency lies with individual organisations, especially private sector ones. In changing their activities or relations to other agents, local organisations have the means to change the appearance of the cluster. At the same time, they have the strongest incentive to make these changes as failing to do so and losing competitiveness can imply the end of their existence.

Second, the study of cluster adaptation also has to highlight the contingencies that agents might face when trying to adapt to change events. These contingencies emerge from the nature of cluster architectures (as outlined in Sect. 4.1), which make actor-driven adaptation a less than straightforward process. Two main reasons are responsible for this. First, the existence of (partly non-linear) agglomeration externalities implies that changes in agent activity can produce unexpected aggregate effects. The sum of individual adaptation thus does not equal cluster adaptation. Second, the cluster as the repository of these externalities has no formal authority over the activities of local agents. While agents' shared understanding of acceptable business practice offers an orientation for agent strategy selection, this mechanism does not work as strongly as formal authority in an integrated organisation. As such, the cluster influences the adaptation process of its actors but does not control it like a formal authority could.

In a way, the adaptation of a cluster to change events could therefore be understood as a special case of decentralised problem solving (by local actors) under externalities at the cluster level. In consequence, the success of adaptation depends on:

- The 'accuracy' of individual firm response (i.e. how well can firms recognise a change event and take proper adjustment measures), and
- The forces at the cluster-level influencing both the available individual activities as well as the link between individual action and collective or aggregate results, i.e. the nature and extent of interdependence between agent activities.

Both points can be addressed from different perspectives. The issue of how well individual firms adapt to change has been advanced in the theory of the firm in the context of organisational inertia (Nelson and Winter 1982). This line of thought argues that firms develop *institutionalised behavioural patterns* in the form of capabilities and routines throughout their evolution to cope with environmental risk and uncertainty. These behavioural patterns can then be hard to unlearn in the face of change events, thereby limiting firm adaptability (see among many others Amin and Cohendet 2000; Levinthal and March 1993; Flier et al. 2003; Greenwood and Hinings 1996; Kondra and Hinings 1998; Schoenberger 1997; Wells 2001; Wicks 2001). As a consequence, a trade-off between competitiveness in one set of environmental conditions under strong routines and capabilities, and adaptiveness of firms to changing environments emerges. However, the conditions under which firms become more or less inert as well as the 'optimal' degree of inertia with regard to competitiveness and adaptability are still not fully understood.

Explaining cluster adaptation from this perspective, i.e. basing it on individual firm responses alone would mean that success depends on the number of 'adaptive' and 'non-adaptive' firms in the cluster. This supposition in turn might lead to an endless regress in describing how a cluster came to host particularly adaptive firms in the first place. The answer to this question would probably lie with coincidences, aspects in its history or in a relationship between cluster features and the resulting adaptiveness of its firms. It is however hard to think of any link between the two latter aspects: Even if a cluster exhibits strong local rivalry and thereby forces its firms to adapt, the kind of industrial change under investigation here significantly changes the basis of competition for cluster firms. While it might be reasonable to assume that firms accustomed to stiff competition are also more alert and receptive to these wider changes, this hypothesis alone carries too little ground to build an entire theoretic framework upon. Most importantly, however, an exclusive focus on organisational adaptation implies neglecting the cluster level interdependences and behavioural contingencies faced by its agents. Relating cluster adaptation to the quality of individual responses alone thus denies those very features (externalities, mode of co-ordination) that underlie the success of clusters in the existing literature. While it is acknowledged that the adaptiveness of individual firms does play an important part for cluster adaptation, their individual behaviour will not be at the heart of the investigation conducted here. Instead, it is treated as an aggregate factor, i.e. firms are treated as representative actors.[55] The focus of this thesis will then lie with the role of architectural, cluster-level factors in steering actor-driven adaptation to good collective outcomes. This neglect of individual organisation's micro-dynamics obviously means that an analysis of the factors put forth in this study tells only part of the story of how agents in clusters can adapt to external events.

The avenue towards understanding the cluster's role in actor-driven adaptation pursued here thus focuses on the second point: The architectural, cluster-level factors that shape the relationship between individual activities and collective outcomes, i.e. the extent of interdependence between the success of individual activities on the one and the behavioural restraints exerted by the mode of co-ordination in the cluster on the other hand. Both aspects can crucially affect the success of individual actor activities. Put differently, the best individual response to change can be suppressed if the collective is not 'ready' for it. Examples would be situations where the adaptive move of one agent will only be successful in the presence of specific activities by other agents (interdependence). A second case would be adaptive moves of an organisation that depart from the established 'rules of the game' and thus induce punishment by other local actors (co-ordination). The latter of these two aspects (agent-level inertia due to the local 'rules of the game') has received extensive theoretic attention in the field of institutional economics. As was argued before (Sect. 3.2.1), both formal and informal institutions arise to facilitate transactions between agents by restricting the range of possible behaviour towards desirable alternatives. Once established, they are difficult to change. This institutional stickiness can stem from the costs associated with institutional change, which is likely to hold in particular for formal institutions (e.g. political bodies or laws). In the case of informal institutions, it is furthermore possible that they "*are reproduced because individuals often cannot even conceive of appropriate alternatives (or because they regard as unrealistic the alternatives they imagine*" (Powell and DiMaggio 1991, p. 11).[56]

As was found in the theory of the firm, very strong and tailored (formal/ informal) institutions in clusters can thus be very effective in a specific environmental setting as they significantly lower transaction cost. Changes in the environment could however imply that previously desirable activities are no longer the best solution. Of course, (informal) institutions usually exhibit some degree of flexibility as agents faced with bounded rationality and uncertainty regarding future developments can never account for all future behaviours.[57] Still, a short-term fixed and very strong local institutional infrastructure has a more constraining effect on in-

---

[55] A similar approach is adapted in the recent actor-based models of Industrial Districts. In them, technology adoption or the creation of local institutions is studied using different general 'rules' for individual behaviour. See Boero et al. 2004; Squazzoni and Boero 2002.

[56] This notion is built on the argument that (informal) institutional arrangements and individual preferences reciprocally influence each other. People within a specific institutional arrangement might then be unable to form diverging preferences that would lead to a change in the former.

[57] This aspect is even truer if (informal) institutions are not products of conscious design but rather as an emerging feature due to agent experiences in interactions.

dividual behaviour and might thus hamper adaptation to change if the strategies required in the new environmental setting constitute a breach with the established understanding of the rules of the game (for accounts of the role of (informal) institutions in organisational change, see among many others Casper 2000; Coriat and Dosi 2002; Flier et al. 2003; Grabher 1993; Nelson and Winter 1982; Powell and DiMaggio 1991; Washington and Ventresca 2004 or Zeitz et al. 1999).[58] Institutional economics thus also arrives at a trade-off between an institution's efficiency within an environmental context and its flexibility in changing surroundings.

The first architectural aspect (agent dependence on other's activities) relates to the notion of agglomeration externalities.[59] Changes in the activities of a sufficient number of agents can come to alter their nature and extent. If many agents decided to cease their research and development or personnel training activities, this would have an effect on knowledge externalities and the availability of skilled personnel in the cluster. In a way, adaptive moves by a number of agents might thus come to reduce the positive effects from agglomeration. At the same time, a reduction of congestion cost due to shakeout effects in the local industry population would be possible. As both externalities depend on the activities of many agents, an individual actor cannot preview the aggregate effects of his individual activities.

What emerges from the reasoning presented above is a duality of drivers for adaptation in clusters. On the one hand, agency resides clearly with individual organisations implying a strong role for the quality of individual replies in shaping the outcome of adaptation processes. On the other hand, the nature of clusters (agent interdependence and co-ordination) means that individual adaptation does not add up to cluster adaptation. As a result, this study focuses on those architectural factors at the cluster level that steer individual activity to good collective outcomes. Put differently, the argument proposed here is that *individual activities 'drive', while cluster-level properties 'steer' adaptation to external events*. Holding the driving factors equal, the study focuses on the *steering qualities* of different cluster architectures. The research thus highlights what structures at the cluster level are more or less conducive to adaptation, if agents behave in a certain way.

4.2.2 Reality bites: Success factors in adaptation

To assist the search for architectural factors in a cluster that can steer agent adaptation to good collective outcomes, this section investigates the findings on success factors to cluster survival in the empirical literature. It does so by addressing

---

[58] The role of formal and informal institutions for change and inertia has been highlighted at different levels of analysis: Individual organisations, regional transformation and at the level of the national political regime. See also Powell and DiMaggio 1991.

[59] The effect of inter-firm networks on organisational performance is investigated in a more general context by economic sociologists. They argue that embeddedness of firms in business and social networks affects individual strategy choices (e.g. Burt 1992, 2001; Boje and Whetten 1981; Fehr and Fischbacher 2002; Granovetter 1985; Uzzi 1997a, 1997b). The notion advanced here is somewhat different in the sense that interdependence with other agents does not affect strategy *choice* but strategy *outcomes*.

a set of questions. How can developments at the local or industrial level turn the one-time advantages of being in the cluster into liabilities? How have existing clusters responded to such changes in their competitive environment? Finally, the review aims at deriving whether clusters that successfully adapted to change events share certain architectural properties that could be used as factors to be investigated in a theoretic model.

Empirical research on clusters and change has found that the majority of events leading to cluster extinction are external to it.[60] Such external challenges can come in many guises. In the past, they were mainly attributed to changes in transport and communication costs. *"Every cheapening of the means of communication, every new facility for the free interchange of ideas between distant places alters the action of the forces which tend to localize industries"* (Marshall 1920, p. 273). Beyond these, however, numerous events can come as challenges to an existing local industry (see Porter 1990 or Ritschl 1927, pp. 843-844 for detail), including:

- technological discontinuities, i.e. new products/ production technologies;
- ubiquitisation of once location-specific inputs (raw materials);
- changes in quality and quantity of demand;
- changes in input requirements shifting production towards sites with better endowments;
- improved or cheaper transportation increasing competition between different clusters in the industry;
- qualitative and quantitative changes in local labour supply (e.g. due to migration);
- changes in transaction cost facilitating a geographic spread of different units of one firm[61] as well as
- politically induced changes in local production cost (e.g. decreasing taxes or capital costs).

The brief and non-exhaustive list of possible external change events already highlights that their extent and their repercussions on clusters will differ strongly. Some events constitute examples of *bifurcations* in the sense of Schumpeterian dynamics of creative destruction altering the entire competitive environment of the industry (Schumpeter 1934b). Others are cases of more or less drastic *perturbations* in parts of the industry. Both industrial bifurcations and perturbations can challenge a cluster to the extent that they might represent a tipping point at which the area's development experiences a *local bifurcation*, i.e. significant short-term changes in economic prosperity with more long-term consequences for the host region's industrial mix. Nonetheless, the response of the cluster to both types of

---

[60] Internal sources of exhaustion include developments specific to the cluster, which make locating in an area less advantageous. These could lie with a loss of local rivalry, the emergence of internal or regulatory inflexibilities (Porter 1990) as well as increasing opportunistic behaviour (Maillat 1998b).

[61] Labelled as a decrease in leadership advantages (*"Führungsvorteile"*) from the spatial concentration of firm activity (Ritschl 1927, p. 843).

external events will differ: Bifurcations at the industry level constitute a cause for more drastic adaptive moves than perturbations.

With respect to the question of adaptation in clusters, different perspectives emerge depending on the timeframe and level of analysis. One view regards the notion of *adaptation through arbitrage* directed at the long-term development of an entire region. This kind of adaptation involves the more immediate intersectoral compensation, i.e. the substitution of losses in economic activity within a cluster hit by an adverse change event through gains in another sector. This avenue of adaptation is obviously only open to areas with a diverse industrial mix. It also encompasses the more long-term issue of industrial substitution where a new industry takes root on the ruins of an old one. This phenomenon is explained by the local presence of important incubators to new sectors such as universities or surviving old industries that can provide inputs to (e.g. resources, infrastructure) or act as customers of firms in newly emerging sectors. In many instances, an old local industry has thus come to form the basis for a new emerging one, i.e. it assisted the region's structural change.[62] For industrial substitution to occur, however, an existing local industry must have survived any environmental change.

The notion of old industry survival directly regards the second perspective on cluster adaptation to change, namely *adaptation through self-organisation*. Research in this vein investigates how and when agents in a local industry challenged by adverse events can adapt and accommodate for these changes. With respect to the empirical evidence, strong differences exist between clusters. In some cases, local industry actors have adapted successfully and retrieved (part of) their former position in the sector while in other instances, the very same change event has lead to decline.[63] Theoretic and empirical research investigating the success factors in surviving clusters has yielded a set of important insights where three factors are stated repeatedly as being decisive for adaptation: Internationalisation, division of labour and mode of co-ordination.

With respect to internationalisation, several authors highlight the role of organisational linkages beyond the cluster's boundaries to help agents adapt (e.g. Maillat 1998a; Bresnahan et al. 2001; Scott 1998; Burt 1992; Grabher 1993; Kern 1996). However, little explanation has usually been provided on the exact causality between international linkages and cluster adaptiveness.[64] Usually, two arguments are put forth:

- International ties help organisations observe more radically different practices conducted outside the cluster. This assists their pursuit and evaluation of more drastic strategy changes.

---

[62] For empirical evidence on this aspect see Braunerhjelm et al. 2000 or Bresnahan et al. 2001. A more theoretic perspective is adopted by Lane 2002 and Swann 1996.

[63] See Braunerhjelm et al. 2000; Bresnahan and Malerba 1999; Cappellin 2003; Cooke 1997; Dalum et al. 2002; Gertler 1996; Grabher 1993; Isaksen 2003; Lee et al. 2000; Saxenian 1994 and Schamp 2000.

[64] An exception being Barthelt et al. 2002 and Cowan and Jonard 2003.

- Exchanges with parties outside the cluster limit the emergence of a very strong and highly specific institutional infrastructure in the cluster, which would act as a strong source of inertia in the face of change events.

This understanding of internationalisation views the increase in cluster adaptability as attributable to the characteristics of individual agents rather than architectural aspects at the level of the cluster. It therefore lies outside the scope of the analysis conducted here.

A more interesting set of findings for this study arises from the striking regularities in the *architectural* properties of clusters that have adapted successfully to a changing environment. The primary focus of work in this vein has been on Italian Industrial Districts (IIDs). Even when the concept was developed, there was some concern that their structure with its predominance of small firms would hamper adjustment to change. It was argued that potentially painful adaptive measures would be hard to co-ordinate in a district composed of many independent actors as a *critical mass* of agents would have to be motivated to make any necessary change. Sebastiano Brusco (ibid 1982) already highlighted the role of collective institutions as a means to overcome co-ordination problems in small firm IIDs.[65] These concerns, while relatively neglected in the time of IID prosperity, came full circle when many districts began to face competitive pressures (Belussi 1999; Bellandi 1996; Bianchi 1998; Guerrieri and Pietrobelli 2000; Paniccia 2002; Russo 1985 and Tiberi Vipraio 1996). The latter stemmed from two sources. Firms in developing countries increasingly addressed some of the traditional markets of IIDs (clothing or shoes). In addition, new production technologies enabled large multinational firms to match the IIDs' capacities in product customisation – in many cases at a lower price. In a sense, changes in competition thus rendered the previously core competence of IIDs more ubiquitous.

Interestingly enough, studies have found a recurring pattern in the adaptation process of small firm IIDs (see Boschma and Lambooy 2002; Cainelli and Zoboli 2004b; Lombardi 2003; Maggioni 2004; Belussi et al. 2003; Guerrieri and Pietrobelli 2001). Beyond the more business-oriented responses of increasing product development effort or cost cutting through internationalisation, regularity could be observed in changes of the way in which co-ordination between agents was carried out. While IIDs were initially a prime example of egalitarian networks of small, independent firms (hopefully) adhering to an informal local culture, their adaptation to change involved a move away from this structure towards an increased hierarchisation of local actors. The latter emerged through the rise of *leader firms* created by a concentration of power (through internal growth, mergers and/ or acquisitions) or through a process of *alliancing* where several independent district firms joined forces in more closely linked business groups. Both developments can be viewed as a way to overcome both the resource constraints faced by small firms as well as the aforementioned co-ordination problem in adap-

---

[65] Brusco 1986 proposes a 'Mark I' and 'Mark II' model of IID development. In the Mark I case, IIDs evolve without any collective institutions. The Mark II scenario stresses the supporting role of collective or public institutions to district firms, e.g. by providing information on international markets or technological developments.

tation with many independent agents. Needless to say that they have spurred considerable concern both with researchers and policy actors since the increased hierarchisation of agents in industrial districts is viewed as a loss of the archetypal form that many held so dear.

A second observation is put forth by the case study comparing adaptation in two computing clusters, the Silicon Valley and the Boston 128 after the advent of the microcomputer (Saxenian 1994). Here, the argument goes that the small, networked firms in the Silicon Valley could better address the newly emerging heterogeneous demand for low-price, customised small systems (Bresnahan and Malerba 1999). This flexibility is usually attributed to their greater division of labour: The more integrated firms of Boston's Route 128 encountered severe difficulties in adapting, partially due to high sunk costs forestalling a significant change in business practice. As a consequence, this specific change event came to favour the Silicon Valley because its structure and functioning worked to produce results that were rewarded by the new demand.

Both examples suggest a trade-off between 'inertia' due to small firm size and co-ordination problems on the one hand and the inflexibility attributable to greater sunk costs associated with changes in large enterprises on the other hand. In other words, the empirical findings point at a link between two cluster-level factors (division of labour, mode of co-ordination) and the success of their adaptation to change events. However, a general causality cannot be derived. While incredibly beneficial and rich in evidence on specific issues, case studies are very bound to their context. Moreover, the complexities involved with adaptation in existing clusters (firm responses, policy influences, nature and extent of change events, etc.) imply that a multitude of factors could underlie the observed results. While empirical research does provide evidence and intuitions on a role for the division of labour and the way cluster agents co-ordinate their activities in adaptation, these results are not ready for generalisation. In order to single out the influence of both aspects, a more theoretic approach is needed that can be tested in a controlled environment limiting influencing factors. Two interesting questions thus emerge from the empirical work and will lie at the heart of the theoretic analysis.

1. What is the role of the cluster's division of labour for adaptiveness? It will be argued that the division of labour influences cluster structure (in terms of number and size of actors) as well as the extent of interdependence incurred by agglomeration externalities. Both aspects impact on cluster adaptation.
2. How does the mode of co-ordination in clusters matter for adaptive performance? It is proposed that the mode of co-ordination influences the goals underlying agent strategy selection. Actors in clusters with a leader firm are likely to exhibit different goals than those operating in egalitarian networks as they have to account for the interests of the leader firm. Differing modes of co-ordination will thus generate different adaptation processes.

In order to answer both questions, a model has to be constructed that can link the performance of agent adjustment to the influence of different cluster architectures. The following sections outline the nature (4.2.3) and the requirements (4.3) for such a model in more detail.

4.2.3 Agents, interdependence, adjustment: Cluster self-organisation

Having established that adaptation in clusters proceeds through the interplay of individual strategies and cluster-level factors while exhibiting interesting empirical regularities, a second important issue emerges. The latter has significant repercussions on the nature of the phenomenon investigated since it regards the time-horizon for adaptation. This matters for adaptation in several ways. First, organisational strategies can differ strongly depending on the timeframe given.[66] Second, at the level of clusters, longer time horizons imply a greater potential for change both at the level of agents (entry, exit, range of activities conducted) as well as with respect to cluster-level factors: While agglomeration externalities and the shared understanding of acceptable business practice characterising clusters might be difficult to modify in the short run, a more long-term perspective offers the opportunity of gradual changes in both, as actors alter their interactions with other agents and renegotiate acceptable behaviour over time. Third, extending the timeframe of analysis can easily imply that instead of studying adaptation in individual clusters, one begins to look at the overall development of entire areas, where issues related to adaptation by arbitrage coming into play.

As a result, one has to ask what the time-horizon for adaptation as it is understood here should be. The importance of this question also stems from the fact that different timeframes of analysis require different modelling tools (see also Fig. 4.3.). Longer timeframes further imply that a greater number of influencing factors can enter the analysis. With respect to both aspects – time and complexity – modelling techniques thus exhibit a *"hierarchy, where increased 'simplicity' of understanding and predictive capability is obtained by making increasingly strong assumptions, which may be increasingly unbelievable"* (Allen 1997, p. 3).[67]

When thinking of different modelling techniques, the first and most abstract one is a description of reality providing *"all detail of everything, everywhere, as well as all perceptions and all points of view"* (Allen 1997, p. 4). For the scientific study of a specific problem, this is obviously of limited value as no regularities or rules can be identified. Any scientific approach therefore requires a more structured perspective. The first abstraction from reality proceeds by classifying real objects according to the similarities and differences between them. A listing of these objects alongside their dates of emergence, development and (possibly) disappearance then yields the evolutionary tree of a phenomenon. For instance, if the problem to be investigated were the economic development of an area over time, these objects would be the relevant local actors. The latter could be classified according to different criteria (industry, size, position in the value chain, etc.). Adding the dates of their emergence and possibly disappearance would then lead to the evolutionary tree showing how economic activity in the locale has evolved up until a chosen instant. This kind of model would show the sequence of birth, change

---

[66] To highlight this aspect, simply consider the difference between strategic and operative activities advanced in the business literature.

[67] The following reasoning is based on Allen 1997 (Chapter 1).

and death of different local industries. As a result, the model would operate with changing elements and thereby changing economic structures in the area.

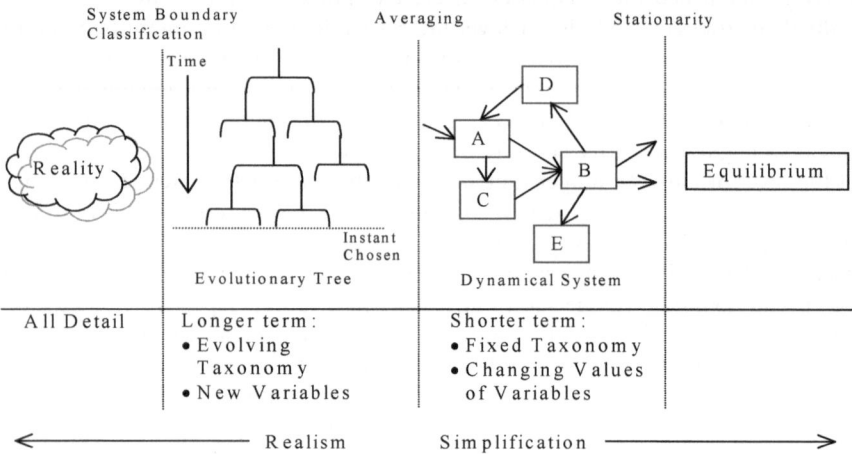

Fig. 4.3. Studying cluster development – Timeframe, assumptions and tools
Source: Allen 1997, p. 4 (with permission from Gordon and Breach Science Publishers)

As mentioned before (Chap. 1) this evolutionary tree of an area's development is characterised by branches representing periods of relative stability and tipping points. In the context of area development, the branches represent a (more or less prosperous) local industry. The tipping points represent external events constituting bifurcations to the development of a specific cluster. These local bifurcations can be the result of industry-wide bifurcations or smaller perturbations in the cluster's environment, both challenging the current practice in the area. The difference between the two types of change lies with the scale and scope of their effects: Industry-wide bifurcations in the Schumpeterian sense challenge any local cluster involved in the sector. Perturbations in turn can affect selected clusters only (e.g. policy initiatives affecting all clusters in the jurisdiction of the initiator). The behaviour of the local industry in the face of a local bifurcation then determines, which development path is chosen after the event. With respect to the economic prosperity of an area, these paths might differ strongly. The success of an existing industry in adapting to external events challenging the cluster can therefore have both short- and long-term repercussions for regional prosperity.

Within the more short-term perspective focussing on one cluster facing an external shock, local agents and their interrelations are constant. As highlighted in the previous two sections, a cluster with a specific architecture (interdependence, co-ordination) and configuration will have evolved by the time the change event occurs. This implies that both model elements as well as their interrelations are fixed. By analysing the immediate response of an existing cluster to change, this study therefore focuses on the processes of self-organisation between cluster agents that follow an external shock. As such, the model is not evolutionary in a stricter sense since it does not allow for changing 'taxonomies', i.e. there are no mechanisms accounting for the emergence of new and disappearance of existing agents, changes the range of agent activities, alterations of the nature and direction of interdependencies resulting from agglomeration externalities nor modifications in the way agents co-ordinate their activities. Put into the perspective of the evolutionary tree, the model therefore investigates the immediate response of agents to a bifurcation event in cluster development.

The focus on this more short term phenomenon can be justified by two arguments. The immediate response of cluster agents to an external event can easily decide, which of the many possible development trajectories following a local bifurcation is 'chosen', i.e. the immediate response can be decisive for the question of decline or adaptation. As a result, it also has also more long-term repercussions on the development of an entire area: Survival of a local industry after changes in its environment not only determines the immediate economic prosperity of the cluster but might also lie at the heart of future structural change through industrial substitution (see 4.2.2). Moreover, investigating the factors mattering for short-term adaptation in clusters can also be justified by the fact that at this point in time, the reversal of 'bad' development trajectories is still relatively easy. Once a given trajectory is established (e.g. an industry starts to decline), the amount of resources needed to change it increases hugely with time. While important for cluster development, success in this first phase of adaptation does not guarantee a similar outcome in later stages. Therefore, there will be *no one to one* relationship between the findings of this study and cluster survival. Instead, results are more to be viewed as factors making successful cluster adjustment and survival more likely. To investigate these decisive factors *influencing* the long-term development of an area, the modelling framework will have to account for a number of aspects.

4.3 The nature of cluster adaptation and resulting model requirements

The previous sections have highlighted that clusters exhibit two peculiarities making their adaptation to change a less than straightforward process. Individual activities are influenced by cluster-level factors, either because certain business practices are not acceptable or because the outcome of a given activity depends on externalities conveyed by the actions of other agents. Cluster adaptation is therefore driven by individual activity and while cluster-level factors can steer the direction of these actions, there is no central decision-making authority. Second, a brief review of cluster's adaptation to change revealed that within the more short-term context of existing industry adjustment, interesting regularities with respect to the relevant cluster-level factors could be observed. In comparative case studies, evidence and intuitions for a role of the extent of division of labour in clusters as well as the way activities are co-ordinated between agents were repeatedly emphasised. This short-term response could moreover also affect the more long-term development of the cluster and its host area: In many instances, a surviving local industry later acted as the basis for newly emerging sectors, thereby assisting an area's long-term structural change through industry substitution. However, the context-bound and complex nature of case study evidence implies that the causalities between cluster-level factors and adaptiveness are 'disturbed' by a multitude of other influencing variables. In consequence, this study proposes the development of a theoretical framework able to explain and investigate the role of these case-study derived architectural factors for cluster adaptation.

Thanks to the nature of the phenomenon to be investigated (the relatively short-term adaptation of an existing local industry), the model to be derived falls into the category of 'dynamical systems'. As the interest of the study lies with in the process of clusters adapting to a change in their environment, equilibrium models comparing two steady states are ill suited to address it. Within the timeframe of the analysis conducted here, the nature of cluster agents, their interdependence due to agglomeration externalities and the way they co-ordinate activities are 'fixed taxonomies'. In other words, while the external event can change the performance of clusters' interdependencies and co-ordination mechanisms (e.g. decrease the competitive value of specific information spillovers), the nature of both will remain unchanged in the short term. What is to be determined then is how agents adapting within the given cluster architecture regarding externalities (interdependence) and co-ordination fare, given that differences in both aspects may exist between clusters. This task involves relating the activities of individual actors to collective outcomes at the cluster level given a specific architecture and configuration of the latter. One way to do this lies with the study of multi-layered systems and their behaviour, which will be explored in more detail in the following chapter.

As the previous theoretical and empirical evidence has highlighted, cluster level factors can matter for the effectiveness of actor-driven adaptation. A model investigating their role thus has to account for a number of issues. First, it needs to mirror the *nature of clusters* as entities composed of multiple agents occupying different stages of the value chain. Second, a link between agent activities, cluster architecture (interdependence, co-ordination) and results in a specific environment has to be established both for individual agents as well as for the collective, i.e. the cluster as a whole. Put differently, a *measure of success* is required. Third, the *two different kinds of change* in the environment (Schumpeterian industry bifurcations versus smaller scale perturbations) affecting the success of the cluster and its agents need to be accounted for. Fourth, the model has to allow for *multiple outcomes of adaptation* processes. Finally, in order to investigate the role of cluster-level factors in adaptation, *differences in cluster architecture* regarding their degree of division of labour (extent of interdependence between agents) and mode of co-ordination (nature of goals underlying strategy choice) have to be modelled. To investigate their effects, a comparison of different cluster architectures regarding adaptive performance can then be conducted. The following chapter shows that these requirements can be met if a more generic understanding of clusters as systems composed of several complex, co-evolving sub-systems (made up of actors at specific stages of the local industry's value chain) is adopted.

5 Modelling adaptation in clusters – The promise of complexity theory

As was outlined in the previous section, any formal model analysing the role of the division of labour and co-ordination for cluster adaptiveness has to meet a number of requirements. First, it has to link the behaviour of cluster agents with the interdependencies caused by agglomeration externalities and the effect of different co-ordination mechanisms provided by the local culture. The resulting outcomes at the cluster level then have to allow for a measure of success for different cluster configurations. Second, the model would need to include the impact brought upon the cluster by bifurcation or perturbation events in its environment. Third, an investigation of the role of the division of labour and co-ordination requires that the model be able to mimic different cluster architectures with respect to both factors. If these aspects are included, the role of division of labour and co-ordination for cluster survival can be investigated by a comparison of adaptive performance for different cluster types.

The very nature of clusters as entities composed of firms and other local organisations whose activities are linked by agglomeration externalities and steered by a co-ordination mechanism implies a relatively straightforward role for the division of labour and the mode of co-ordination in their adaptation. Due to agglomeration externalities, the success of individual firm activities hinges on the activities of others. As a result, the likelihood of successful adaptation for each cluster depends to some extent on the degree of interdependence between firm activities. Interdependence is linked to the division of labour in the cluster: The smaller the number of activities performed by individual firms, the greater their dependence on those of others will tend to be. With a growing dependence on others, the importance of co-ordination between agents will increase as well. Co-ordination can internalise some of the interdependencies into agent decision-making thus achieving better overall results. As can be expected, the effectiveness of any co-ordination mechanism will therefore be based on the extent to which it captures cross-agent interdependence. This co-ordination can then be sustained in two ways: Either through 'voluntary' behavioural restraints exerted by equally powerful agents or by differing distributions of authority between cluster firms.

This section highlights that the means for studying the role of both factors in cluster adaptation are available if one moves towards an understanding of clusters as complex, adaptive systems. Sect. 5.1 provides an overview of systems and complexity theory in order to derive whether clusters match that definition. It is resolved that it is possible to understand clusters as complex adaptive systems if agency does not have to reside at the system level. Sect. 5.2 then shows that the modelling requirements outlined so far:

- agent-driven, interrelated dynamics (5.2.1),
- success measures for specific cluster configurations (5.2.2),
- the impact of environmental bifurcations or perturbations (5.2.3) and
- different degrees of division of labour and modes of co-ordination (5.2.4),

can be accounted for in a formal model of clusters when basing it on Kauffman's N/K methodology. The concluding section in this chapter outlines the aspects that need to be resolved for building an N/K based simulation model of clusters. They involve arriving at an understanding of the different parameters, relating the dynamics between the levels of analysis (agents, groups and the cluster), mirroring different extents of environmental change as well as finding suitable proxies for the extent of division of labour and modes of co-ordination. Prior to developing an N/K based simulation model (Chap. 7), the factors shaping the behaviour of N/K systems in general will be reviewed in more detail (Chap. 6).

5.1 Clusters as complex adaptive systems

Building a case for understanding clusters as complex adaptive systems requires a definition of systems, complexity as well as complex, adaptive systems. Sect. 5.1.1 therefore outlines the related fields of systems and complexity theory showing that the latter provides a more realistic account of the characteristics found in many social systems. Sect. 5.1.2 then reviews whether the features found with clusters (Chap. 4) match the requirements for complex adaptive systems. It is found that both concepts coincide if agency is allowed to reside at the level of system elements (agents) rather than in the system itself. Having established that clusters can be understood as complex adaptive systems, Sect. 5.2 then introduces a suitable modelling tool able to account for the requirements outlined before.

5.1.1 Systems and complexity theory: An overview

Scientific analysis in its first step proceeds by structuring reality through a classification of real objects according to their similarities and differences in chosen characteristics (see also 4.2.3). These objects are frequently made-up of smaller sub-units, i.e. they constitute aggregations or collections of smaller entities. Some of these aggregations can be referred to as *systems*. They exist at various scales: Molecules consist of atoms, organisms are comprised of cells or – in the context described here – clusters encompass different local organisations. As a result, the level of analysis adopted determines what a system is and what constitutes an element. Put differently, elements at one level of analysis can be systems at another, lower level of investigation. For instance, organisations constitute elements in the system cluster but at the same time can be viewed as systems of individuals when adopting an organisation-level perspective. Similarly, clusters can constitute system elements when higher-level spatial structures like regions or nations are investigated (Wilson 2000). Many definitions have been put forth to outline what distinguishes systems from mere aggregations of elements. They centre on four characteristics (Skyttner 1996, pp. 35-36):

1. Element interdependence: The behaviour of each element has an effect on the behaviour of the whole as well as on the behaviour of other elements;
2. Organisation: A system is a structure that has organized components;
3. Distinctness: A system is anything unitary enough to deserve a name; and
4. Purpose: The system's design is aimed at performing some function.[68]

Combining these features, one possible definition would be the following: "A system is a set of interacting units or elements that form an integrated whole intended to perform some function, [..., i.e.] any structure that exhibits order, pattern and purpose [, ... as well as] some constancy over time" (Skyttner 1996, p. 35). The Systems Theory emerging in the 1950s thus took up on a well-established observation by recognising that many distinguishable phenomena in life are made up of smaller, interrelated units.

The dynamics of any system can then be derived from its internal laws regarding element behaviour and interdependence as well as environmental influences impacting on the system. However, the laws of motion influencing system behaviour differ greatly between system types. In decomposable systems where an understanding of the behaviour of elements suffices to understand the whole, dynamics are mechanic in the sense that they can be explained by *"'causal mechanisms', where components influence each other and form systems, in which the change seen in one part is explained by change in another, or in the external environment in which the system is embedded. These mechanisms could be written down as mathematical equations, as mathematical models expressing fundamental laws of nature, and then used to predict behaviour."* (Allen 1997, p.1).

This reductionist perspective of explaining the whole through the sum of its parts was long predominant not only in the natural but also in the social sciences. Describing social systems as mechanical entities was however strongly at odds with everyday experience. People as the most elementary units in any social system can respond, react, learn and change differently, depending on their respective experiences and personalities. This implies that existing social systems would exhibit a greater variety of element behaviour than could be accounted for by a set of fundamental laws governing individual's actions.[69] A second problem with the mechanical view on social systems regards the possibility of interdependence between individuals. If the attainment of goals pursued by a rational individual

---

[68] As a matter in fact, there is some degree of dispute regarding the extent to which systems have to be directed at a purpose. This issue will play a fundamental role in assessing the 'system' nature of clusters in Sect. 5.1.2.

[69] This plurality of possible actor activities was usually accounted for by generating representative individuals like the 'homo oeconomicus' that made decisions according to predefined procedures (e.g. rational choice under full information).

hinges on decisions taken by others, even full information of each agent on the ways in which his own activities contribute to attaining a specific goal does not ensure an optimal decision. The agent choosing a certain action would still depend on the activities of others. In many instances, agents also act in a *simultaneous-move game* where they cannot preview what others are going to do. This implies that, there is a possibility that agents do not reach their goals at one point in time because others acted in an unforeseen way. Most social systems are thus more like worlds *"where many players are all adapting to each other and where the emerging future is extremely hard to predict"* (Axelrod and Cohen 1999, p. xi).

While the reductionist perspective failed to account for many characteristics of real-world social systems, adopting a more realistic view also implied that it became increasingly difficult to advise agents or designers on the right behaviour, e.g. regarding policy interventions in systems. Recent research in the natural sciences has however bred hope that the particularities of social systems are not fully at odds with predictive capacity: Even complex, non-decomposable systems of many different elements interacting with one another in intricate ways are able to derive ordered macro-level behaviour from their microdynamics. In the natural sciences, such *emergent properties* were found in experiments where molecular interaction gave rise to chemical clocks or macro-structures in heated liquids.[70] In the social sciences, the idea of obtaining ordered and non-obvious macro-dynamics from relatively simple behavioural rules for system elements is prominent in the work on emerging segregated neighbourhoods (Schelling 1978). Understanding how the macro-level behaviours arise can then help predict the future development of such systems in changing circumstances at least to some extent, although prediction in complex systems lies more with a description of trends than with the deterministic outcomes found in decomposable, mechanical ones (Nijkamp and Reggiani 1998).

The science of such *complex* system dynamics that are on the edge between orderly and chaotic behaviour was born with the introduction of thermodynamics.[71] In contrast to gravity in the Newtonian paradigm, heat did not act upon innate matter but transformed it (*"Ignis mutat res"*; Prigogine and Stengers 1984, p. 103). The system, its elements and thereby its behaviour could thus change over time, depending on environmental conditions. As a result, in contrast to the Newtonian frame of reference, time became a crucial element of analysis and a system had to be defined *"... not as in the case of dynamics, by the position and velocity of its constituents ... but by a set of macroscopic parameters [...]. In addition, we [had] to take into account the boundary conditions that describe the relation of the system to its environment"* (Prigogine and Stengers 1984, p. 105-106).

---

[70] Chemical clocks are reactions in liquids where the concentration of different chemicals changes in regular intervals, colouring the mixture one way or another as relative concentrations are altered (Prigogine and Stengers 1984). Heated liquids in turn can display stable hexagonal structures when the rates of rising hot molecules is equal to that of descending colder ones at given temperature and pressure conditions (Allen 1997).

[71] In their interesting and comprehensive review, Prigogine and Stengers 1984 label the introduction of Fourier's law of heat propagation in solids (in 1811) as the birth date of complexity science. Later work has been centred in the Santa Fe Institute in New Mexico.

While it was found that closed thermodynamic systems always evolved towards thermodynamic equilibrium (i.e. maximum molecular entropy), orderly behaviour emerged, when systems were open to exchanges with their environment and exhibited some degree of non-linearity in the interactions of their elements. Within these kinds of systems, orderly macrobehaviour became possible due to the emergence of system-level properties, i.e. the whole system became more than the sum of its parts. As a result, higher system levels could no longer be fully understood by studying lower ones. This is attributable to the fact that emergent properties (Byrne 1998, p. 4) are caused by and influence element behaviour. They can therefore not be explained by analysing lower-level dynamics alone. Such emergent properties in non-decomposable systems can stem from a variety of sources:

- Non-linear interdependence
  Non-linearities within a system can apply to the nature of interactions of system elements: If the joint contribution of element A and B yields an effect greater than A+B, a study of either element will not tell the whole story.
- Multiple contingent interaction
  A similar notion applies if the relationship between two variables A and B is altered by the state of a third variable C (Byrne 1998, p. 2). Dissecting the system and studying either of these variables without taking their interdependence into account would miss part of overall system dynamics.
- Reciprocal interdependence
  Any study of elements will only lead to a full description of the behaviour found at the system level if the effects of element dynamics are linearly additive and operate in one direction, i.e. if the behaviour of the system is determined by that of its parts. If the system however influences the behaviour of its elements, a reductionist perspective breaks down.
- Environmental events
  In open systems where environmental events can impact on the (determinants of the) behaviour of system elements, a stochastic component emerges implying that the system's development trajectory is no longer deterministic but depends on more or less likely environmental events.

As a consequence, the structural evolution of non-decomposable systems proceeds as a dialogue between their average behaviour in periods of stability and fluctuations around these values. In some instances, fluctuations can lead to the emergence of a local bifurcation in the development path of the system where the future development becomes hard to predict and small events can set it off into very different directions. Studying the system at a moment of disturbance is furthermore of interest since the outcomes of bifurcations in the development of natural systems are often also time-irreversible since the information required to reverse the events leading to a bifurcation is infinite (the so-called '*entropy barrier*'; Prigogine and Stengers 1984, p. 295).

As a result, periods of determinism and chance (at bifurcations) co-exist within such complex systems. Both aspects (deterministic and chance behaviour) are shaped by the value of control parameters inside and outside the system, which is why the term 'complex adaptive systems' is sometimes used to describe that they develop in response to internal and external pressures. As has been outlined before, this notion very closely corresponds to the observable dynamics in social systems. In consequence, there was a feel that complexity theory offered a great promise to the social sciences by providing both a better methodological grasp on the characteristics of real social systems while still obtaining ordered developments that could be used for prediction. *"We urgently need to revive systems approaches to the social sciences, and the complexity/ chaos programme provides us with a way of doing this which overcomes the very real difficulties encountered when the models of systems available to us were equilibric or at best close to equilibric. Far from equilibric systems are very different indeed"* (Byrne 1998, pp. 8-9). Applications of complexity theory have thus grown in many areas of the social sciences including organisation studies and various forms of spatial analysis.[72] If clusters were complex (adaptive) systems and if their laws of motion could be derived, we could attempt an analysis of what internal factors are crucial for success or failure in adaptation once a cluster is hit by an environmental change. The first question to be addressed in this context however involves determining whether clusters can be understood as complex adaptive systems.

5.1.2 Are clusters complex adaptive systems? The issue of agency

Defining complex systems relates both to the general definition of systems outlined in the previous section and to a notion of what constitutes 'complexity' in them. To date, there is no agreement on a definition or measurement of complexity (Axelrod and Cohen 1999, p. 15). In this study, a definition is adopted where complexity is the length of a concise description of an entity's regularities and random features. Any object's complexity is therefore context-dependent in the sense that the length of any description hinges on the level of detail chosen, a common language and the knowledge and understanding of the world: *"[I]magine that you are entering the village of a group of hitherto uncontacted Indians in the Brazilian jungle, and that you know their language from experience with another group speaking the same tongue. Now explain to your hosts the meaning of a tax-managed mutual fund. It should take quite a while"* (Gell-Mann 2002, p. 15).

This notion of complexity can apply to two different aspects of the system: Structure and dynamics. *"Although complexity is a notion connected with dynamic systems, the term 'complexity' is also often used to describe a certain degree of 'complicatedness' in an ordered system's structure (static complexity). [This kind of complexity arises in system configurations] where the components of a system are put together in an intricate and interrelated way"* (Nijkamp and Reggiani 1998, p. 133). Static complexity thus relates to the number of hierarchical levels in the system, the variety of its' elements and the (strength of) connectivity.

---

[72] Examples include the evolution of cities and regions (Allen 1997; Garnsey 1998; Pumain 2004), development of transportation networks, structural changes in the industrial mix of areas and the likes. For an overview of complexity theories in the spatial sciences see Nijkamp and Reggiani 1998.

At the same time, complexity can be associated with the behaviour of the system (dynamic complexity). Complex system dynamics are driven by control parameters within and outside the system[73] as well as the laws of motion describing the system's behaviour for specific parameter sets. This behaviour can be complex in the sense that it is difficult to predict, as there may be multiple possible outcomes. This complexity in system dynamics can originate both for systems with low and high static complexity in their structures. For instance, the Feigenbaum sequence describes the behaviour of many systems in which there is "*a periodicity to bifurcation which doubles with each successive bifurcation. This represents a route from simple determination through a realm of complexity within which there are multiple but limited outcome situations towards a realm of chaos in which there are very large possible sets of outcomes*" (Byrne 1998, p. 22; Prigogine and Stengers 1984, pp. 126 and 169).[74]

A definition of complex systems would therefore relate to establishing the necessary characteristics of a system as well as the features relating to complexity in its structure and/ or behaviour.[75] Regarding the system nature of clusters, Sect. 4.1 has already established that they are characterised by different elements (local agents) that are interdependent due to the cluster-level effects of agglomeration externalities and the mode of co-ordination. Moreover, clusters have been argued to evolve into a distinguishable configuration of agents in a certain industry and territory.[76] As a consequence, they meet the first three requirements of systems (agents, interdependence, and distinctness). It is the fourth one, 'purpose', that is problematic. Do clusters exhibit purpose in their own right?

The question of purpose is in itself debated, i.e. there are other definitions of (complex) systems in which the aspect of purpose is not a necessary requirement. Some (Gell-Mann 2002) propose that complex, adaptive systems have to be able to actively respond to changes in their environment to meet the definition. In this understanding, a complex adaptive system would have to exhibit some level of agency at the system level. Others (Holland 2002; Axelrod and Cohen 1999; Allen 1997) have argued that complex adaptive systems consist of (potentially several layers of) interacting components (agents) with non-linear relationships that are open to environmental influences. In addition, adaptation (changing behaviour and

---

[73] The values of inside and outside parameters apply both for closed and open systems (the latter are able to exchange energy, matter or information with their environment). In case of open systems, the influence of the environment is however *crucial*.

[74] In most instances, the distinction between the different regimes is made according to the existence and nature of attractors for the system. The latter can be stable points, periodic or oscillating as well as strange attractors. The latter are usually illustrated by the Lorenz attractor or the 'butterfly effect' where changes in initial system conditions as small as a butterfly batting its wings can produce very different outcomes (see Byrne 1998, p. 19).

[75] It has been noted before that for systems to display complex dynamic behaviour without constant outside perturbations, a minimum degree of static complexity is needed (Prigogine and Stengers 1984, pp. 298-299)

[76] Steps towards a complexity perspective on clusters have been undertaken by Boero et al. 2004; Curzio and Fortis 2002 (especially Rullani 2002 therein); Fornahl and Brenner 2003; Garnsey 1998; Karlsson and Johansson 2005 or Squazzoni and Boero 2002.

learning) has to be possible at the agent level. As has been established in Sect. 4.1, agency in clusters lies with the agent rather than the cluster-level.[77] Since the architectural properties of clusters derived in the previous chapter significantly impact on individual agent behaviour and thereby adaptation of the collective of agents residing in the cluster, it is argued here that clusters exhibit *indirect agency*. While the cluster cannot act itself to pursue a certain adaptation strategy, it still steers individual activity to better or worse results from a system perspective. As a result, adaptation cannot be understood by studying the level of agents alone. One can therefore argue that clusters meet the second definition of complex adaptive systems, which will be adopted for the remainder of the study: The influence of system-level properties on agent activities imply that it makes sense to study cluster adaptation to change as an example of decentralised problem solving (with agency residing at the level of individual actors) under interdependency at the system (cluster) level.

Research in the natural sciences has yielded a number of methods to study the behaviour of complex adaptive systems (Nijkamp and Reggiani 1998). These range from models with non-linear dynamics, neural network theory, and chaos theory up to different kinds of descriptions of the dynamics of macro structures based on interlocking micro-level elements. Such agent or element based models of complex systems exist for a variety of methodologies and in very different contexts like game and network theory to name but a few (see Lux et al. 2005; Schweizer 1997 or Weidlich 2000 among many others). Another approach originally developed in biology is the N/K model proposed by Kauffman 1993. As will be shown in the remainder of this and the following chapter, the N/K methodology constitutes a promising avenue for investigating the adaptation of clusters to change events since it can easily accommodate the modelling requirements outlined in Sect. 4.3. The details of how this is achieved will be elaborated on in the following section. It can however be said already that the N/K model enables a comparative analysis of the adaptive performance of differently configured complex systems, i.e. it can highlight, what cluster architectures perform better or worse in adapting to changes in their environment.

5.2 Agents, interdependence and fitness: Introducing the N/K model

As was mentioned before, a variety of tools are available for the modelling of complex (adaptive) systems based on their microdynamics. Some of these proceed by assuming an average interdependence of system elements where the exact distribution of linkages is not important. In other words, in such averaging models of complex systems, it is of lesser importance which elements are connected

---

[77] This is not to imply that collective activities are impossible. However, these would have to be initiated by one or a number of local agents. Agency, even in this context, is associated with the element but not the system-level.

(Boccura 2004). For clusters, such an approach would mean that it would not matter, which firms were linked by externalities. As argued in Chap. 4, interdependencies thanks to agglomeration externalities are a key feature of clusters and it is thus intuitive that their distribution will matter. Due to agglomeration externalities, the activities of specific cluster agents are interrelated. This matters for the outcome of agent-driven adaptation. The following sections therefore review one model of complex systems that can operate with directed interdependencies: Kauffman's N/K model (1993). While N/K is not the only possible avenue to modelling clusters understood as complex adaptive systems, it constitutes the most suitable methodology for the study conducted here. This is attributable to the fact that the nature of N/K systems allows for an integration of all aforementioned modelling requirements (Sect. 4.3) in a very straightforward way.

First, N/K enables an analysis of complex systems characterised by directed interdependencies between its elements. These interdependent elements can be within the control of one agent (K) or under the jurisdiction of different agents (C), i.e. the model allows for intra and inter-agent interdependencies. Especially the latter (the cross-agent externalities measured by C) are important in the analysis of clusters as they mirror the effects of agglomeration externalities. Second, N/K provides a measure of success for different cluster configurations through the concept of fitness landscapes. Third, the effects of bifurcations and perturbations in the cluster's environment can be accounted for by altering these very landscapes to differing degrees. Fourth and most importantly, the factors shaping the dynamics of N/K systems (structure, search, selection) can be used to build different cluster architectures with respect to the two stylised empirical realities observed.

As a result, by creating and comparing clusters with different extents of division of labour and distinct modes of co-ordination, their adaptation to both bifurcations and environmental perturbation can be modelled and compared allowing for a great variety of measurable cluster behaviour while maintaining a limited number of parameters. As a result, the N/K framework will constitute the basis for the simulation model to be developed in the following chapters.

5.2.1 The importance of directed interdependencies

Originally developed to study the evolution of a population of genes, the N/K model can be extended to a variety of complex, interdependent systems since it describes the latter in very general terms by two variables: The number of system elements (N) and the degree of interdependence between them. Each of the N system elements can take on a discrete number of states (An). The possibility space of the system (S) then contains all possible system configurations (s) for given values of N and A_n (Eq. 5.1).

$$s = \{a_1; a_2; a_3; ....; a_n\} \text{ with } s \in S \qquad (5.1)$$

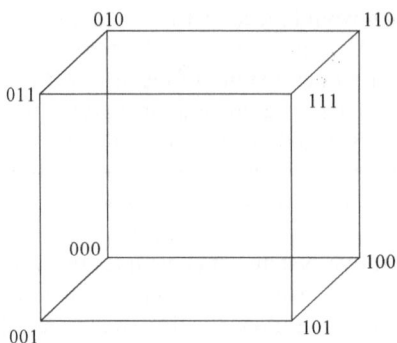

Fig. 5.1. Possible system configurations for N=3, A_N=2

A frequent case is A_n=2, i.e. the N elements can take on two values (usually [0; 1]). This yields a possibility space of 2^N system configurations. As an example, in a system consisting of N=3 elements with A_n=2=[0;1], there are 2^3=8 possible configurations, one of which is {0,0,0} (see also Fig. 5.1.). In consequence, N/K allows for a multitude of possible system configurations even with a limited number of elements. These configurations can all reflect different system behaviours as they correspond to alternative combinations of element states. This kind of multiple dynamics would be hard to achieve in other models like those representing a system in differential equations.

5.2.2 Measuring success by fitness landscapes

Depending on the environmental conditions, the state of an element then contributes to the *fitness* (the likelihood of survival) of the system (Kauffman, 1993, p. 33). In other words, each element state (e.g. n_i=0 and n_i=1) has a fitness value w_{na}, which is drawn randomly from a uniform distribution between 0 and 1 (e.g. w_{10}=w(n_i=0)=0.4 and w_{11}=w(n_i=1)=0.3). The environmental conditions determine the value of a specific element's state for the fitness of the system. For instance, if the system described by N and K is the genome of a plant, one element could be an allele enabling survival with relatively little water. This capability would contribute more to the fitness of the system 'plant' in an environment where water is scarce. The fitness of the entire system (W(s)) is then determined as the mean value of the element fitness contributions (Eq. 5.2).[78]

$$W(s) = \frac{1}{N} \cdot \sum_{n=1}^{N} w_n(s_n)$$

(5.2)

---

[78] Depending on the nature of the system, this assumption may not be realistic. In the context of technology adoption, Frenken and Nuvolari 2003 suggest a weighting of fitness contributions to reflect different user preferences for specific technology characteristics.

This formulation allows for a mapping of all possible system configurations and their relative fitness in *fitness landscapes* (Wright 1931). These fitness landscapes consist of the fitness values W of all possible system configurations (s). In them, "*each attribute (gene) of an organism is represented by a dimension of the space and a final dimension indicates the fitness level of the organism*" (Levinthal 1997, p. 935).

	n_1	n_2	n_3	$w_1 (a_1=0) = 0.5$	$w_2 (a_2=0) = 0.1$	$w_3 (a_3=0) = 0.7$
w_1	-			$w_1 (a_1=1) = 0.6$	$w_2 (a_2=1) = 0.3$	$w_3 (a_3=1) = 0.3$
w_2		-				
w_3			-			

$\{a_1,a_2,a_3\}$	w_1	w_2	w_3	W
$\{0,0,0\}$	0.5	0.1	0.7	0.43
$\{0,0,1\}$	0.5	0.1	0.3	0.30
$\{0,1,0\}$	0.5	0.3	0.7	0.50
$\{1,0,0\}$	0.6	0.1	0.7	0.47
$\{0,1,1\}$	0.5	0.3	0.3	0.37
$\{1,1,0\}$	*0.6*	*0.3*	*0.7*	*0.53**
$\{1,0,1\}$	0.6	0.1	0.3	0.33
$\{1,1,1\}$	0.6	0.3	0.3	0.37

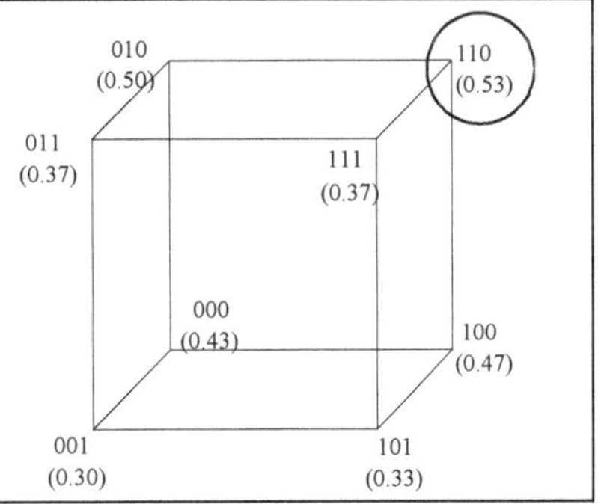

* system optimum

Fig. 5.2. System configurations and fitness levels for N=3 and K=0

	n_1	n_2	n_3	$w_1(a_1=0 \cap a_3=0)=0.7$	$w_2(a_2=0 \cap a_1=0)=0.1$	$w_3(a_3=0 \cap a_1=0)=0.5$
w_1	-		x	$w_1(a_1=0 \cap a_3=1)=0.6$	$w_2(a_2=0 \cap a_1=1)=0.1$	$w_3(a_3=0 \cap a_1=1)=0.4$
w_2	x	-		$w_1(a_1=1 \cap a_3=0)=0.3$	$w_2(a_2=1 \cap a_1=0)=0.2$	$w_3(a_3=1 \cap a_1=0)=0.4$
w_3	x		-	$w_1(a_1=1 \cap a_3=1)=0.4$	$w_2(a_2=1 \cap a_1=1)=0.2$	$w_3(a_3=1 \cap a_1=1)=0.5$

$\{a_1,a_2,a_3\}$	w_1	w_2	w_3	W
$\{0,0,0\}$	0.7	0.1	0.5	0.43
$\{0,0,1\}$	0.6	0.1	0.4	0.37
$\{0,1,0\}$	*0.7*	*0.2*	*0.5*	*0.47**
$\{1,0,0\}$	0.3	0.1	0.4	0.27
$\{0,1,1\}$	0.6	0.2	0.4	0.40
$\{1,1,0\}$	0.3	0.2	0.4	0.30
$\{1,0,1\}$	0.4	0.1	0.5	0.33
$\{1,1,1\}$	0.4	0.2	0.5	0.37

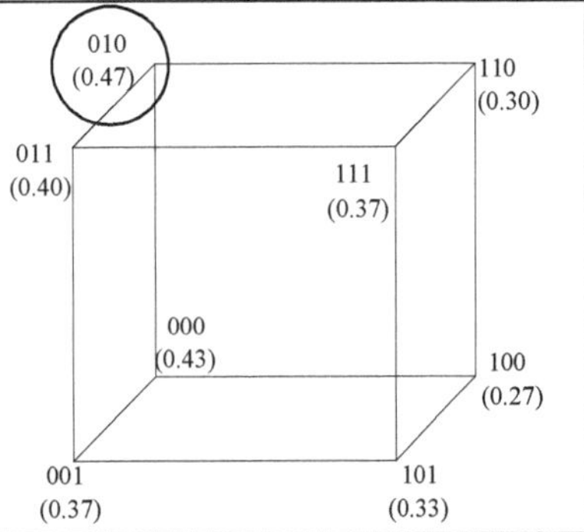

* system optimum

Fig. 5.3. System configurations and fitness levels for N=3 and K=1[79]

---

[79] Here, interdependence exists between n_1/n_3, n_2/n_1 and n_3/n_1, i.e. we arrive at 'conditional' fitness contributions of each element state, given the state of one other element.

	n_1	n_2	n_3
w_1	-	x	x
w_2	x	-	x
w_3	x	x	-

$w_1(a_1=0 \cap a_2=0 \cap a_3=0)=0.7$ $w_1(a_1=1 \cap a_2=0 \cap a_3=0)=0.6$
$w_1(a_1=0 \cap a_2=1 \cap a_3=0)=0.6$ $w_1(a_1=1 \cap a_2=1 \cap a_3=0)=0.3$
$w_1(a_1=0 \cap a_2=0 \cap a_3=1)=0.7$ $w_1(a_1=1 \cap a_2=0 \cap a_3=1)=0.3$
$w_1(a_1=0 \cap a_2=1 \cap a_3=1)=0.6$ $w_1(a_1=1 \cap a_2=1 \cap a_3=1)=0.6$
$w_2(a_2=0 \cap a_1=0 \cap a_3=0)=0.3$ $w_2(a_2=1 \cap a_1=0 \cap a_3=0)=0.4$
$w_2(a_2=0 \cap a_1=1 \cap a_3=0)=0.2$ $w_2(a_2=1 \cap a_1=1 \cap a_3=0)=0.2$
$w_2(a_2=0 \cap a_1=0 \cap a_3=1)=0.1$ $w_2(a_2=1 \cap a_1=0 \cap a_3=1)=0.2$
$w_2(a_2=0 \cap a_1=1 \cap a_3=1)=0.3$ $w_2(a_2=1 \cap a_1=1 \cap a_3=1)=0.4$
$w_3(a_3=0 \cap a_1=0 \cap a_2=0)=0.4$ $w_3(a_3=1 \cap a_1=0 \cap a_2=0)=0.6$
$w_3(a_3=0 \cap a_1=1 \cap a_2=0)=0.4$ $w_3(a_3=1 \cap a_1=1 \cap a_2=0)=0.4$
$w_3(a_3=0 \cap a_1=0 \cap a_2=1)=0.5$ $w_3(a_3=1 \cap a_1=0 \cap a_2=1)=0.4$
$w_3(a_3=0 \cap a_1=1 \cap a_2=1)=0.5$ $w_3(a_3=1 \cap a_1=1 \cap a_2=1)=0.5$

$\{a_1,a_2,a_3\}$	w_1	w_2	w_3	W
$\{0,0,0\}$	0.7	0.3	0.4	0.47
$\{0,0,1\}$	0.7	0.1	0.6	0.47
{0,1,0}	*0.6*	*0.4*	*0.5*	*0.50**
$\{1,0,0\}$	0.6	0.2	0.4	0.40
$\{0,1,1\}$	0.6	0.2	0.4	0.40
$\{1,1,0\}$	0.3	0.2	0.5	0.33
$\{1,0,1\}$	0.3	0.3	0.4	0.33
{1,1,1}	*0.6*	*0.4*	*0.5*	*0.50**

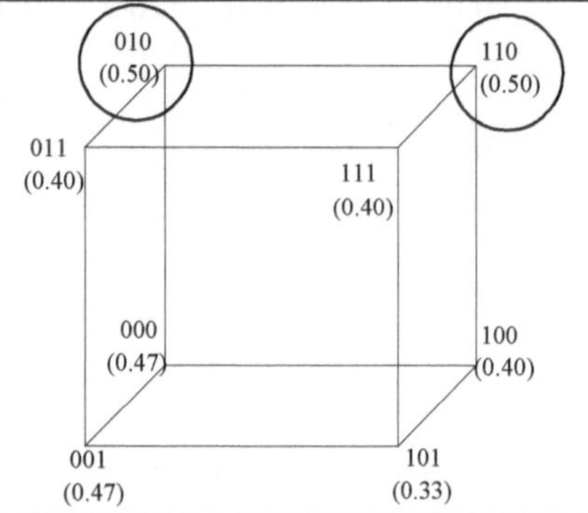

* system optima

Fig. 5.4. System configurations and fitness levels for N=3 and K=2[80]

The shape of a system's fitness landscape depends on two factors: The randomly drawn fitness values for element states (w_{na}) and the degree to which interdependencies between elements exist, i.e. the extent to which the fitness value of an element does not only depend on its own state, but also on those of other elements. The value of K illustrates the degree of interdependence. If K=0, the fitness value of all elements only depends on their respective state. If K>0, the fitness value of any element's state in the system is affected by the states of K other elements, i.e. each element has A^{K+1} different fitness contributions for different combinations of element states (Kauffman 1993, p. 239). In the most extreme case K=N-1, the fitness value of each element depends on the states of all other elements. Figs. 5.2. to 5.4. illustrate the fitness landscapes for an independent (K=0), interdependent (K=1) and fully interdependent (K=2) system with N=3 and A_n=2=[0;1]. It shows that the number of system optima increases with *complexity*, i.e. with interdependencies between system elements.

The nature of interdependencies is conditional on the kind of system analysed. In the case of *co-evolving systems* where elements are controlled by different agents, interdependence can occur at the level of the agent or between agents. The

---

[80] The fitness contribution of an element depends on its own state and that of all others.

first aspect – intra-agent interdependencies – reflects the phenomenon that activities might be interrelated at the level of the firm. For example, strategy choices in research, production and marketing are interdependent in the sense that firms need to align them to produce good results. If the system to described consists of one entity controlling all system elements (e.g. a firm controlling all aspects of its strategy: Rivkin 2000; Rivkin and Siggelkow 2003 or innovations: Fleming and Sorenson 2001; Frenken 2001), all interdependencies are of the intra-agent nature measured by the parameter K. The second aspect, inter-agent interdependence is a characteristic of systems where several agents control different system elements. They are measured by a third parameter, C. In both cases, the values of K and C indicate the average number of other elements influencing the fitness value of any system element. Put differently, if K=3, the fitness of each system element on average depends on the states of three other elements within the control of the same agent. If C=3, the fitness of each system element depends on three elements that are controlled by other agents. The distribution of both types of interdependencies can however be uneven in the sense that some elements may be more 'central' in the system (exhibit higher levels of interdependence) than others.[81]

As has been shown elsewhere (see Kauffman 1993), systems characterised by greater interdependency (or complexity), i.e. greater values of K(C) tend to exhibit more 'rugged' fitness landscapes in the sense that both multiple optima are possible and that the fitness values of similar system configurations tend to differ more strongly. If K(or C)=0, the fitness value of all elements only depends on their respective state. Two system configurations differing by one element state would thus have relatively similar fitness values (differing by the difference in fitness of that element's two states). If K>0, the fitness value of any element's state is affected by the states of other elements and vice versa. As a result, system configurations differing by the state of one element may differ substantially in fitness if that element has a lot of repercussions on the fitness of other elements. Due to a similar effect, the average fitness of optima decreases as element interdependence increases (the *complexity catastrophe*; Kauffman 1993, p. 52). The existence of interdependencies implies that there may be conflicting constraints in the optimisation of elements, i.e. an improvement in fitness for one element might come at the expense of another. This means that the average fitness of optima in systems characterised by conflicting constraints tends to be lower than in independent systems where each variable can be optimised individually.[82] The system structure (in terms of N, K and C) therefore determines the structure of the fitness landscape.

---

[81] In the classical N/K system, interdependencies between elements are distributed randomly. As a result, as long as K<N-1, some elements of the system can be independent while others receive a larger share of links.

[82] The system structure determines how "rugged" or "multi-peaked" the fitness landscape is and how distinct the different optima turn out to be. The exact fitness value of different optima however depends on the random allocation of (conditional) fitness values to the respective element states. In consequence, although average fitness of optima decreases with system complexity, some optima in high K systems may out-compete those of low K systems (Hovhannisian and Valente 2004). The importance of this aspect for the simulation results will be elaborated on in more detail in chapter 7.

5.2.3 Bifurcations or perturbations in the fitness landscape

In the N/K(C) model, a drastic change in the environment would result in a change in fitness contribution of elements. As was mentioned before, environmental conditions decide on the fitness value of element states. Returning to the example of the plant's genome, the fitness contribution of an allele enabling survival in environments with little water will be reduced in the face of a climate change leading to an abundance of the formerly scarce resource. Environmental changes would therefore change the fitness landscape of the system, which can happen in two different ways:

- The fitness contributions of the system's elements can be altered while leaving the structure of their interdependencies intact (Levinthal, 1997, 2000).
- The structure of interdependencies between elements is redistributed. As a result, the fitness contributions of those elements affected by the redistribution are altered.

In the context of clusters, the former way is a more realistic perspective. A change in the cluster's competitive environment does not necessarily alter the structure or the effects of agglomeration externalities. For instance, the Italian Industrial Districts did not lose the agglomeration effects enabling them to produce highly customised products in a very flexible way. Instead, the value of these effects in the new competitive environment was altered. Being able to produce varying quantities of customised products became less valuable as integrated multinational producers became increasingly able to do so (Sect. 4.2.2). In many other instances, the impact of external changes on clusters did not imply that they became bad at what they were doing. Instead, the results of their way of organising business were less valued by the environment once the change had occurred. As a result, it will be argued here that environmental change affects the fitness values of system elements while leaving the structure of interdependencies intact.

As was mentioned in the previous chapter, changes in the cluster's environment can come in many guises with differing impact. Some of them constitute smaller perturbations, such as temporary fluctuations in demand quantity or smaller changes in consumer preferences. Others can come as bifurcations to the industry in a Schumpeterian sense implying that they alter the entire competitive environment. Instances like this could be the introduction of radically new products or even production technologies. In many instances, Schumpeterian dynamics do not mean an instantaneous change of industries in the sense that both old and new industry prevail for a limited time (McGahan 2004). Based on this notion, the argument proposed here is that existing clusters do stand a chance in adapting to Schumpeterian dynamics in their industry, even through this adaptation may be a lot more difficult to achieve than in the case of environmental perturbations.

The difference between bifurcations in a Schumpeterian sense and smaller environmental perturbations can then be found in the range of the elements affected by the change event. It will be argued further (Chap. 7) that drastic changes, i.e. bifurcations, affect the entire landscape. As such, they are represented by altering the fitness contribution of all elements. A perturbation in the system's environment can in turn be accounted for by changing the fitness contribution for a limited number of elements. Such an event would imply that only a part of the fitness landscape is deformed. The question of how the system can adapt to these changes then relates to its ability to find new configurations that exhibit a high fitness in the altered landscape. This is a direct consequence of the dynamics of N/K systems which are affected by their *architecture*, i.e. the conditions shaping system *structure, search and selection*.

5.2.4 Structure, search, selection: Division of labour, co-ordination

The dynamics of N/K systems relate to their ability to improve upon the fitness of an initial configuration $s_{init}=\{a_1, \ldots , a_n\}$. This increase in fitness is obtained through trial and error search activity where the system *searches* its fitness landscape for superior configurations by modifying elements. It *selects* these modifications if they provide a higher fitness value. If the new fitness value is equal or lower than that of the previous configuration, the system rejects the new configuration. This strife for increased fitness corresponds to a 'walk' on the fitness landscape towards better configurations. A configuration is stabilised if it has the highest fitness level attainable given the system's initial configuration and its' underlying dynamics.

As will be shown in more detail in Chap. 6, the dynamics of N/K systems are driven by a trinity of factors, namely *structure, search and selection*. Changes in either of these have significant effects for the configurations the system can encounter and thereby also for its fitness in a given environment. As a result, structure, search and selection also matter for the system's adaptability to change events – both with respect to bifurcations as well as perturbations. The system's ability to walk on rugged fitness landscapes influences its propensity to find better configurations once these landscapes have been changed by environmental events.

Chaps. 6 and 7 will therefore elaborate on the role of structure, search and selection for adaptability first in a general N/K sense (Chap. 6) and in the specific context of clusters (Chap. 7). In the latter setting, it emerges that the extent of division of labour shapes the structure of the cluster's fitness landscape by determining the degree of interdependence between elements (firm activities). The mode of co-ordination in turn is argued to influence the selection mechanism employed by cluster agents, i.e. it determines which fitness level has to be improved for a modification to be considered successful. Comparing the adaptive performance of clusters characterised by different fitness landscape structures and distinct selection mechanisms will then allow for a more conclusive analysis of the role played by both factors in cluster adaptiveness.

5.3 Clusters as N/K systems: Parameter definition and system dynamics

As has been highlighted in Sect. 5.1, the nature of clusters as multi-layered entities where individual activities leads to system-level properties that feed back on agent behaviour is best understood when adapting a complexity perspective. Furthermore, it was found that out of the many modelling tools available, Kauffman's N/K model allows for an accommodation of the modelling requirements outlined in Chap. 4 in a very intuitive and straightforward way. Thanks to its generality, N/K avoids an over-parameterisation (Frenken n.a.): The effects of agglomeration externalities are accounted for in a very general way by introducing positive and negative interdependencies between system elements. This conceptualisation avoids a quantification of the exact strength of both types of externalities which would be very hard to achieve as is witnessed by the debate on the importance of knowledge externalities for firm innovation (e.g. Baptista 2000; Baranes and Tropeano 2003; López-Bazo et al. 2004 versus Beaudry and Breschi 2000; Helsley and Strange 2002).[83] Moreover, environmental change is introduced by changing (parts of) the fitness landscape which is a more general formulation than one obtainable in other models where the extent of the impact caused by different types of change events would have to be quantified. As such, N/K saves a number of parameters that are not relevant to the question investigated here but that would enter a model of cluster adaptation to change if a different methodology were adopted.

The fact that clusters consist of different agents conducting competing and complementary activities in the local value chain however implies that we are not dealing with the integrated one agent systems that constitute much of the body of analysis in the N/K framework. Instead, clusters can be understood as co-evolving systems of different agent populations (henceforth: groups), each controlling a certain range of industry activities. This aspect implies that both the factors governing the behaviour of an agent and the dynamics in his group have to be linked with the development introduced by the co-evolution of several agents and groups (Chap. 6). Applying the N/K model to an understanding of clusters is furthermore going to involve a number of extensions to the original model as social systems differ from biological ones in a number of aspects. For instance, search and selection mechanisms can be more elaborate in social settings where agents consciously decide on the modifications to be executed. The issue of relating individual agent to group and cluster dynamics as well as the extensions to the original N/K model will be discussed in the following chapter before developing the simulation model.

---

[83] Additional work on space and innovation is found with: Baptista 1998; Breschi and Lissoni 2001; Koschatzky 1998; McKelvey et al. 2002; Oerlemans and Meeus 2002; Orlando 2000; Zucker et al. 1994, 1997; Zucker et al. 1998

Part IV Model development – Clusters as complex adaptive systems

> *If I had to model industrial clusters, I would search carefully for mid-dle-level semi-empirical rules that might persist all the way from the real world down to the highly simplified model, and test the model by seeing if it clarifies how the rules arise*

Murray Gell-Mann 2002, p. 24

Having established that clusters can be defined as complex adaptive systems, the following chapters investigate how complexity theory and especially the N/K framework could be used to establish causalities regarding the roles of two empirically derived cluster-level factors (degree of division of labour and mode of co-ordination) for their adaptation to change. As such, the qualitative simulation model developed here aims at finding explanations for the role of both factors and adaptive performance within the N/K framework.

Understanding clusters as co-evolving N/K systems implies that one deals with different agents aggregated into groups that conduct parts of the activities in the local value chain. Interdependence between agent activities can then occur among activities controlled by one agent reflecting that strategy choices (e.g. in research and production) may have to be aligned within firms to produce good results. The effect of agglomeration externalities is mirrored in the existence of cross-agent interdependencies between value chain activities, e.g. situations where the outcome of end-producer activities depends on those undertaken by suppliers.

A role for the degree of division of labour at the vertical level of the cluster as well as the mode of co-ordination between its agents is then found with the effect of both aspects on cluster adaptability, i.e. its likelihood of finding good new configurations after a change event (chapter 7). Arguing that the mode of co-ordination provided by the local rules of the game influences the goals underlying agent strategy selection, the model introduces different forms of indirect inter-agent co-ordination by assuming different selection mechanisms at the agent level.

The extent of division of labour in turn impacts on the degree to which interdependencies between activities exist at the level of agents or between agents. Both aspects matter for the number, cluster-level optimality and spread of modifications found by each agent group, thereby driving cluster adaptability. An investigation into their exact roles is then done by comparing the dynamics regarding agent-driven adaptation within static cluster architectures (part V).

6 Micromotives and macrobehaviour[84] – Dynamics of N/K systems

When analysed from a complexity perspective, clusters become systems with emergent properties that are attributable to the activities of agents and the interdependencies between them. Furthermore, due to their openness to a specific environment, any stable behaviour of clusters can be disrupted by changes in the former. As was established in the previous chapters, an understanding of the dynamics of clusters at the time of a disruption in their development is relevant from a short and a long-term perspective, thanks to the role of small events in deciding on the development trajectory following the change event – as well as the difficulty of reverting a trajectory, once it has been established. Moreover, the N/K model was found to provide a good means to analyse the role of architectural factors in clusters for their adaptation. This is due to its ability to account for both the characteristics of ideal-typical clusters, change and adaptation as well as the stylised empirical regularities discovered in case study research.

In many instances, empirical research investigating the success factors of surviving clusters has pointed out the relevance of the degree of division of labour and the mode of inter-agent co-ordination in them (Saxenian 1994; Cainelli and Zoboli 2004b; Lombardi 2003; Maggioni 2004). The methodological limitations underlying case study research (contextual evidence with multiple possible causes of observed phenomena) however imply that, the exact causalities of whether, when and how division of labour and inter-agent co-ordination matter for cluster adaptation have not yet been derived. Bearing with Murray Gell-Mann's quote (see part IV), the model therefore takes the stylised empirical facts as its point of departure, integrates them into a theoretic framework and attempts to provide an explanation for:

1. When and how can a greater division of labour help cluster adaptation to change? Where and how does the trade-off in flexibility between decentralised networks and integrated organisations materialise?
2. When and how does the mode of co-ordination between cluster actors lead to good results of their joint adaptive effort? Which mode of co-ordination is preferable and why does one observe a shift towards alliance and hierarchy structures in many empirical cases?

When using the N/K framework to investigate why and how the extent of division of labour and the mode of co-ordination in clusters matter for adjusting to change,

---

[84] Quoting Schelling 1978 (book title).

adaptability relates to the system's (the cluster's) capacity of finding better (fitter) configurations after a change event has occurred. The ability of an N/K system to generate improvements in its configuration has been extensively discussed in the literature finding a number of parameters that shape system dynamics. In N/K systems with co-evolving agent populations, these relate to the drivers of and interrelations between the behaviour at the different sub-levels of the system.

At the micro-level of the system (individual agents), the behavioural parameters are agents' *search* and *selection* mechanisms. At the meso-level (agent groups), the link between agent and aggregate group behaviour acts as a key determinant of resulting group dynamics. Deriving the behaviour of the macro-level of the system (cluster) from these micromotives then involves addressing how the different levels of analysis (agent, group and system) are linked to one another. This mainly relates to the *structure* of the system in terms of how much different agent groups are linked by C externalities. As will be explained in more detail in this chapter, the ability of clusters to find improved configurations after a change event depends on three related aspects: The number and system-level optimality of configurations discovered by each agent group (which in turn depends on agent behaviour, group dynamics and the number of agents per group), as well as the spread of changes in one group throughout the system (C).

In studying the literature on the dynamics of co-evolving N/K systems, this section aims at three goals. First, understanding how different behavioural parameters shape the dynamics of N/K systems will be useful to determine where the extent of the division of labour as well as the mode of co-ordination in clusters can be introduced into the N/K framework. Moreover, it highlights how differences in both factors are best accounted for. By linking this aspect to the existing knowledge on parameters and system behaviour, it is also possible to derive hypotheses on why, how and when the extent of division of labour or the mode of inter-agent coordination helps adaptation. Second, the two parameters to be investigated here are not the only factors influencing the behaviour of N/K systems. In consequence, the review provided in this chapter is also meant to assist in making 'good' choices for hidden parameters in the simulation model; i.e. to devise settings for the parameters that will not lie at the heart of analysis in this study that are as neutral as possible for the resulting system behaviour. Third, an understanding of the general model dynamics combined with the features of the simulation model to be provided in the following chapter will assist the analysis of simulation results (Chap. 8).

Prior to going into detail on the simulation model, the upcoming sections therefore start by investigating the determinants of agent (6.1) and group (6.2) behaviour in the N/K framework, building both on Kauffman's original and subsequent model extensions. The third section then links agent, group and system behaviour by investigating the dynamics of co-evolution. Sect. 6.4 summarises the key aspects to be taken into account when developing the simulation model of cluster adaptation to change.

6.1 Agent dynamics: Structure, Search and Selection

A first step in understanding the behaviour of co-evolving N/K (C) systems like clusters regards taking a 'bottom-up' perspective starting with the behaviour of individual agents. As has been highlighted in previous work on the properties of N/K systems, their behaviour crucially depends on a trinity of factors S^3, involving all aspects regarding system *structure* as well as agents' *search* and *selection* mechanisms.

6.1.1 Structure: The nature of fitness landscapes

As was outlined before (Sect. 5.2.1), the system's structure influences its fitness landscape by determining the number of system elements as well as the interdependencies between them. In this context, more complex systems (greater numbers of elements and interdependencies) are often subject to conflicting constraints in their optimisation. As a result, the average fitness of optima for such systems tends to be lower than that of less complex ones. In comparison to entirely interdependent systems, however, some degree of interdependence between elements improves the average fitness of optima, thus leading to an optimum degree of system complexity. This effect is attributable to the randomly drawn fitness contributions for system elements. As the number of links between elements increases, so does the number of conditional fitness contributions for different element states – and thereby the number of random draws: Rather than drawing one fitness contribution for each element state, the fitness values of elements in interdependent systems are redrawn depending on the extent of links between elements. For a limited degree of element interrelatedness, there is a likelihood of drawing higher fitness contributions (as compared to the independent case) and therefore average fitness increases. As the number of interdependencies passes a threshold value, however, the aspect of conflicting constraints in the optimisation of element starts to materialise more and more strongly, enabling the emergence of a *complexity catastrophe* (Kauffman 1993, p. 54).[85]

Subsequent work regarding the structure of fitness landscapes has added two main aspects. In the original model, interdependencies were distributed randomly within the system. It could however be argued that some systems exhibit more regular distributions of interdependencies in the sense that only specific subsets of elements are connected to each other. This understanding of uneven distributions of interdependencies also leads to the notion of decomposability: If interdependencies are unevenly distributed between elements, it is possible that the system can be decomposed into smaller blocks of interdependent elements.

---

[85] Due to a similar notion, very complex systems can exhibit some optima with very high fitness values although the *average* fitness of system optima declines.

	1	2	3
W_1	-	X	
W_2	X	-	
W_3			-

	1	2	3
W_1	-		
W_2	X	-	
W_3		X	-

Fig. 6.1. Decomposable versus non-decomposable N/K systems

Fig. 6.1. shows the two types of N/K systems.[86] In the first of the two, elements one and two are interrelated, while the third element of the system is independent of both. As a result, such a system is decomposable into two parts: {1;2} and {3} which can then be optimised independently. With respect to system performance, optimising these decompositions is equivalent to trying to optimise the system as a whole. In the second system, however, the first element affects the fitness of the second and the second affects the fitness of the third element. Separating this kind of system will always result in some interdependence between parts. If it is nonetheless decomposed (e.g. {1;2} and {3} as before), optimising parts does not necessarily lead to an optimum system configuration since there is an interdependence between elements (two and three) that is not taken into account.

	1	2	3	4	5	6	7	8
W_1	-	X						
W_2	X	-						
W_3			-	X				
W_4			X	-				
W_5			*x*		-	X		
W_6					X	-	*x*	
W_7							-	X
W_8							X	-
	Block 1		Block 2		Block 3		Block 4	

Fig. 6.2 Near decomposable N/K systems

In between these two extremes, there are instances of *near-decomposable* systems (Nelson and Winter 1977; Simon 1981) where the majority of linkages is distributed into blocks of elements and only a few linkages between blocks exist (Fig. 6.2.). In the near-decomposable system depicted here, most interdependencies within four element blocks. Decomposing such a system into four blocks of two elements each would lead to a neglect of two interdependencies, namely between

---

[86] Interdependencies are highlighted by 'x' and their structure can be read following the columns for each system element 1, 2 and 3. For instances of 'x', the system element influences the fitness value (w_n) of another element. For example, element 1 has an influence on w_2, i.e. its state influences the fitness of element two. Each element also influences its own fitness value, which is why the diagonal in both systems exhibits instances of '-'. However, in the N/K framework, only interdependencies between elements are counted which is why the link between each element's state and its own fitness value is not highlighted.

element four and five as well as between element six and seven. In some instances, neglecting some of the interdependencies within a system can come to have benefits regarding the speed of search processes, but it implies a chance of having inferior system configurations as a result (see Sect. 6.1.2). The success of decomposition strategies will thus depend on the structure of the system, i.e. the extent to which interdependencies are distributed to allow for the separation of systems into parts.

A second extension to the N/K model addresses the uni-dimensionality of the fitness function. Put differently, N/K is "*based on the idea that each element in a system performs its 'own' sub-function within the system with regard to the attainment of <u>one</u> overall function on which selection operates*" (Frenken 2001, p. 87, own emphasis). Many systems that can be described in the general N/K framework would however be evaluated according to several criteria. One way to include this aspect is through the introduction of elements affecting different functions that in turn determine system fitness (Altenberg 1994, 1995 and 1997). Interdependencies between elements would then regard the extent to which they affect the same functions (Fig. 6.3.). In the system described below containing three elements and two functions, elements one and two as well as two and three are interdependent since they jointly affect one of the two functions. In the terminology used above, such a system would be non-decomposable.

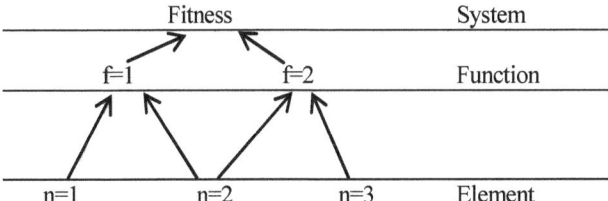

Fig. 6.3. Elements, functions and interdependence
Source: Adapted from Frenken 2001, p. 88 (with permission from the author)

The type of fitness landscapes described by Altenberg can also be used to model evolving landscapes where elements affecting different functions can be introduced as the system develops. In the context investigated here, however, the one-dimensional N/K fitness landscapes with block-distributed interdependencies will be used in order to facilitate analysis while maintaining a reality based account of empirical realities in clusters (for additional detail, see Sect. 7.1.1). The next issue relevant for system performance in differently structured N/K fitness landscapes then depends on how the system can move on its landscape towards higher fitness configurations. Given a particular structure of the landscape, this aspect crucially depends on the search and selection mechanisms employed.

6.1.2 Search: Landscape exploration by mutation

With a randomly assigned initial configuration, a system is placed at a specific spot in its fitness landscape. Depending on how the system can *walk* this landscape, i.e. depending on its search and selection mechanisms, different points are attainable from any given initial configuration. The first factor impacting this walk of the system's fitness landscape is the search process employed, i.e. the number of element states altered as the system seeks better configurations. The most extreme case discussed extensively by Kauffman 1993 refers to *local* search, i.e. random changes of one element state. In this search process, a system compares its fitness level of its current configuration to that of another differing from it in the state of one element. If the other configuration exhibits a higher fitness, the differing element is modified accordingly. In the original model, the system does not search for an overall attainable optimum but chooses the first superior configuration it encounters. For systems with multi-peaked fitness landscapes (i.e. $K \neq 0$), this search strategy exhibits the strongest path dependence in the sense that modifications made in early stages of the processes affect the attainable fitness level.

Local search is thereby most susceptible to becoming locked-in to local optima. As an example, consider a fitness landscape exhibiting two peaks at configurations $s_1\{1;1;1;0;0;0\}$ with $W(s_1)=0.9$ and $s_2\{1;1;1;0;0;0\}$ with $W(s_2)=0.8$. Let a system start with configuration $s_{init}\{0;0;0;0;0;0\}$. Both optima shall be accessible via fitter 'one-mutant' neighbours from this configuration. Since their difference in configuration is such that it affects different elements of the system (e.g. the superior configuration requires changes in only the first three, the inferior one only in the last three elements), the question of which optimum the system uncovers exclusively depends on where it finds the first better one-mutant neighbour. If local search first leads to a change in one of the last three elements, e.g. $s_{init+1}\{0;0;0;0;1;0\}$, the system will walk towards s_2: In order to reach the global optimum s_1, this change would have to be reversed, bringing the system back to its initial, inferior configuration. In systems striving for immediate fitness improvement, such a change would not be made, locking the system into the first local optimum it encounters. If only local search is conducted, the initial system configuration and the search path chosen determine the level of fitness the system can reach. Path dependence in one-bit mutational search can thus lead to lock-in with 'local' optima, i.e. configurations that are not optimal in the entire fitness landscape but constitute the best attainable alternative given the initial configuration and the chosen search path.

At the other side of the spectrum, all elements may be modified in *long-jump* search corresponding to a full system reconfiguration. Again, if the fitness level encountered by the new configuration exceeds that of the previous one, the modifications are retained. Technically, such a search strategy would enable any system to encounter the global optimum, although the probability of locating the one (or few) globally best system configuration(s) exclusively through long-jump search is very low: The likelihood of discovering superior configurations decreases both with the fitness of the current configuration as well as the number of successful long-jumps already conducted (Kauffman 1993, pp. 70 and 74). This is especially pronounced in correlated landscapes (systems with K>0), where the likelihood of successful long-jump search is reduced so strongly that the fitness level attained by systems only conducting long-jump search dwindles towards the mean fitness in the possibility space. Long jump search on such landscapes thus suffers the *complexity catastrophe*, especially as N increases while K remains constant (Kauffman 1993, pp. 72-73).

Between the two extremes of changing only one or all elements, different degrees of *intermediate* search, changing a number of elements between 1 and N are possible. This search strategy is less path-dependent than the one-element mutational search but can still lead to system lock-in at inferior configurations. At the same time, its likelihood for success is higher than for long-jump search and the fitness values encountered tend to be higher. Both factors however decrease as the number of elements modified approaches N. As has been found elsewhere, intermediate search can enable decomposable systems to find a global optimum provided the number of elements modified per search step corresponds to the number of elements in the largest interdependent system part (Frenken et al. 1999; Frenken and Valente 2003; Kauffman et al. 2000; Simon 2002).

From these characteristics, Kauffman 1993, p. 74 arrives at three time scales for search processes: When the initial system configuration has a low fitness value, local search yields variants that are only slightly fitter while long-jump search is still relatively likely to be successful. In the midterm, fitter variants are more readily discovered through local search until a (local) optimum is reached. To leave the latter (in the long run), the system again depends on a successful long-jump search. Depending on the search process employed, the system's initial configuration can thus lie in the *basin of attraction* of one or several (local) optima. The propensity for lock-in is higher, the lower the number of elements modified in the search process.[87]

A second measure of effectiveness for the different search processes (apart from their ability to find the global optimum) is the time it takes to encounter superior system configurations. As the system grows in complexity (N and K), the search space and thus the time required for finding improvements in its configuration increases exponentially. Any system has A^N possible configurations, which can have very different fitness values depending on the degree of complexity. Following Simon 1981, it is therefore argued that the search for superior system configurations can also proceed through a division of labour (Marengo et al. 2000; Frenken 2001), i.e. the system is divided into separate, smaller parts (*decompositions*) that are optimised individually. By splitting the system into smaller units, the search space for each individual unit becomes smaller and enables a faster encounter of superior subsystem configurations. The success of this search strategy with respect to the entire system's fitness then depends on the latter's decomposability as well as the search and selection processes employed within system parts:

---

[87] The issue of landscape structure and optimal search processes is prominent in many applications of the N/K framework to the study of technological innovation. See among many others Auerswald et al. 2000; Fleming and Sorenson 2001; Frenken and Nuvolari 2004 or Rivkin 2000.

The aforementioned issue of search and lock-in to local optima also apply to the respective system decomposition.

The way modifications are selected at the level of decompositions is also important for system fitness: If only strategies enhancing the fitness of the subsystem are selected, then any remaining interdependencies (in the case of non- or only near-decomposable systems) between decompositions would imply a loss of fitness at the system level (for additional detail, see Sect. 6.1.3). In the case of such non-optimal search and selection processes and only near-decomposable systems, decentralised search thus faces a trade-off where greater speed of search (greater decentralisation by forming smaller decompositions) is obtained at the expense of optimality of results from a system perspective.[88] This trade-off can be alleviated in two ways: Finding system decompositions that best reflect its actual structure (i.e. minimising interdependencies between parts) or by *implementing selection criteria that internalise interdependencies between decompositions*.

6.1.3 Selection: Evaluation of modification fitness

As was outlined before, a system walking its fitness landscape usually has a number of possible improvements within the reach of its search process. A second aspect shaping system dynamics therefore regards, which of these improvements gets to be selected. Selection in turn has both a chance and a systematic element. It firstly depends on whether a better configuration is 'found' by the system. Second, selection also relies on the success criteria employed, i.e. the question of when a system judges a modification to be better than its current configuration. Selection criteria thus determine when search is deemed successful.

Sect. 5.2.2 already introduced fitness as the key selection criterion. Modifications in the system's configuration were only chosen, if they improved fitness. Given this criterion, multiple mechanisms remain. The first and most investigated regards the level of fitness to be improved, i.e. is search geared at improving fitness for the entire system, a part of it or simply the element to be modified? The second and less frequently discussed case regards the timing of fitness improve-

---

[88] This issue is prominent in many applications of the N/K methodology to the study of organisations. In a similar vein to the one pursued here, these studies investigate the optimum degree of division of labour and the best organisational incentives in order to maximise firm (system) fitness. The main difference between these approaches and the one pursued here lies with the lack of central authority in the case of clusters which leads to a differential treatment of search and selection at the level of system decompositions (see also Chap. 7). For applications of the N/K framework in an organisational context, see among many others: Dosi et al. 2003; Dosi and Marengo 2003; Ethiraj and Levinthal 2004b; 2004a; Gavetti and Levinthal 2000; Kauffman and Macready 1995; Levinthal 1997, 2000; Levinthal and Warglien 1999; Marengo et al. 2000; Merry 1999; Rivkin and Siggelkow 2003 as well as Siggelkow and Levinthal 2003.

ment, i.e. do agents require immediate rewards or are they willing to accept short term losses in fitness for the sake of long-term, higher fitness values?[89]

Any division of labour in search and system optimisation introduces the question of the criteria for selection. If search is conducted at the level of system decompositions or even individual elements, its' goals might differ. The notion of a selection mechanism captures this idea by determining the relevant fitness level for the retention of modifications. An individual, group or system-level selection mechanism expresses that variations in configuration are chosen if they increase the fitness level of the affected element, the decomposition/ group, or the entire system. The difference between search activity and selection mechanism is that the former corresponds to *action* while the latter describes the *thinking*. For instance, in a setting where each system part can perform search effort individually while the results of search are evaluated at the level of the system, "*action would be 'local', while thinking would be 'global'*" (Dosi et al. 2003, p. 421).

If search activity is conducted in system parts (groups or individual elements) with a corresponding reward scheme (i.e. to increase the fitness level of the subsystem), the complexity of the system is reduced and the speed of finding optima increases (Marengo et al. 2000, p. 771). However, the increase in speed of such a group-wide search process can come at the expense of the optimality of results: The optima reached for groups or even individual elements only add up to a system optimum if there are no interdependencies between subsystems.[90] This implies that a decomposition scheme has to be found which minimises (or even eliminates) interdependencies across subunits. As a result, the optimum configuration obtained through a divisional search and reward process need not be the optimum one from a system perspective.[91] This trade-off is even more pronounced for element-level search and retention mechanisms and moves in the opposite direction as the size of sub-systems increases (i.e. slower but more system-level optimal search results). The same concerns apply if the selection mechanism is based on *satisficing*, i.e. if selection favours fitness increases in a chosen set of elements over that of the system as a whole or that of other system parts (Frenken et al. 1999; Simon 1981). In an opposite direction of argument, selection mechanisms only allowing for pareto-optimal solutions in the sense that the mean fitness of all elements is improved without decreasing that of individual elements will again lower the speed of encountering superior system configurations (Frenken and Valente 2003).

---

[89] This aspect is implicit in the notion of search 'depth' (Hovhannisian and Valente 2004).

[90] If search and reward mechanism exist at the element level, finding an optimum is only possible for K=0.

[91] If the system is decomposed without full knowledge of the nature of its interdependencies, only those elements with the strongest observable connections are grouped together (Marengo et al. 2000, p. 772). The trade-off between search speed and optimality of results implies that agents with the optimum decomposition strategy might not dominate the population because their search takes too long. Agents with a less 'true' decomposition are more successful in the short term and out-compete the former. The population can thus lock-in with local optima (see Marengo et al. 2000; Frenken et al. 1999; Frenken 2001).

So far, the discussion of selection mechanism has been focussed on an identity between the range of elements in search and selection. Put differently, agents search and select modifications if they improve the average fitness of the range of elements they control. Depending on the extent of elements controlled, this can correspond to the entire system or decompositions of it. An interesting additional phenomenon will be shown to emerge if this assumption of the identity of elements operating in search and selection mechanisms is abandoned. As Chap. 7 will highlight in more detail, allowing for elements _outside_ the control of the individual agent to enter her selection criterion mediates this trade-off. It increases the optimality of results from the system perspective while not taking away all the benefits of faster search results in system decompositions. The argument proposed here is that even if agents aim at improving fitness for all N elements, they do so by changing the n elements under their control. This enables a faster encounter of improvements within that decomposition as compared to agents trying to improve the system through search over all N elements (Sect. 7.2.2).

6.2 The dynamics of agent groups

The previous sections have highlighted the importance of fitness landscape structure, search and selection criteria for the behaviour of individual agents. In the case of _integrated_ N/K systems where all elements are controlled by one agent, the points raised before determine agent and system behaviour. In most instances, several integrated N/K systems can then compete within an environment. They may represent entities as distinct as several possible configurations of an organism's genome or differently structured firms. The role played by population dynamics in these cases is limited to birth and selection. In a 'survival of the fittest'-perspective, they allow for insights on the fittest possible system configuration(s) or the system characteristics enabling advantages in the selection game.[92] Put differently, population dynamics are limited to finding best performing instances.

The role of population dynamics is very different if the population under study is a group of agents controlling a decomposition of a higher-level system, i.e. if it forms part of a co-evolving system of different agent groups. Here, population (henceforth: group) dynamics matter for the behaviour of the entire system as they determine how the linkages connecting co-evolving groups are affected by changes in the latter: The greater changes in activity within a group of agents, the stronger (all else held equal) their repercussion on other groups in the system and system performance.[93] The extent to which activity changes within a group is in turn a composite of two factors. Alongside the criteria determining individual agent behaviour outlined in 6.1, the number of agents per group will matter for the number of modifications tried per group in each time interval. Second, the selec-

---

[92] See for instance the studies on competition between organisations with different search processes in Frenken et al. 1999 or Marengo et al. 2000.

[93] As will be shown in Sect. 6.3, this aspect becomes more important when co-evolving agent groups are more strongly linked by C externalities.

tion criterion employed to derive best agent performance at the group level is important. Similarly to the notion advanced in Sects. 6.1.2 and 6.1.3, selection mechanisms aiming for best performance at the level of the group are more likely to find better configurations among its agents as the search space in the group is smaller than in the system as a whole. However, selection mechanisms aiming at maximum system fitness operate more slowly for the sake of better overall results. The extent to which changes are executed at the level of the group and how they improve system fitness will furthermore matter for the stability of the process of co-evolution in systems of interconnected agent groups.

6.3 Co-evolving agent populations: Self organisation and performance

The role of agent and group dynamics for the system's behaviour hinges on whether one is dealing with *integrated systems* where a population of agents controls all system elements and competes for survival or whether the issue under study are disintegrated, *co-evolving systems* where system elements are controlled by several, different agent groups. As was emphasised before (Chap. 4), clusters consist of a horizontal and a vertical dimension. The former encompasses agents conducting competing activities within one stage of the value chain while the latter comprises all agents with complementary activities along the value chain. In addition, interdependencies between firm activities (agglomeration externalities) occur both at the horizontal and the vertical dimension of the cluster. As a consequence, the success of individual firms is shaped not only by their own activities but also by the behaviour of agents in the horizontal and the vertical dimension.

This finding is very much in line with the definition of co-evolution in the N/K model, where the former is defined "*as a process which couples the N/K landscapes of different 'species' such that adaptive moves by one species deform the landscapes of its partners*" (Kauffman 1993, p. 238). Each 'species' or agent group controls a subset of the entire system (ecosystems in Kauffman's original conceptualisation). Elements can then be linked to other elements within the control of individual agents in the group or to elements belonging to agents in another group. Put differently, the fitness contribution of elements (n) depends on the state of the element (A_n) and in some instances on influences of a certain number of states of other elements either within (K) or outside the element's group (C).[94] In a way, this extends the conditional fitness contribution of specific elements (see also 5.2.1) to include the values of element states in another species, meaning that specific elements now have A^{K+1+C} different fitness contributions. Clusters in the understanding provided here therefore constitute a case of co-evolving systems encompassing agent groups that are linked due to agglomeration externalities. In such co-evolving systems, a number of interesting peculiarities emerge depending

---

[94] C measures the number of each species' elements whose fitness contribution is affected by the states of elements in another species.

on the nature of inter-system interdependence (C) and the dynamics in each sub-system (agent group).

The agents controlling the elements that are linked beyond their group can be viewed as to be playing a game where they choose a strategy (an element state of 0 or 1), which can have differing fitness *payoffs* as a consequence of the strategies chosen by agents in other groups of the system. The question of stability in the development of co-evolving complex systems then relates to the issue of whether self-organisation of agent groups leads to *Nash-Equilibria* where the state chosen for elements affecting the fitness landscape of other groups remains the same. This aspect depends on three factors: The payoff structure for different strategy combinations, the assumptions underlying the way the game is played by agent groups (myopic versus non-myopic strategy choices) and the aggregate behaviour of agents in the different groups. The process of self-organisation within and between agent groups can then allow for two outcomes: Either the interacting species keep deforming each others' fitness landscapes as they develop or the system moves towards *stability*, i.e. a steady state at which the (local) optimum of each group is consistent with that of the other interacting ones (Kauffman 1993, p. 245). The stability of co-evolutionary processes then first relates to the assumptions about the way the game regarding cross-population (C) interdependent elements is played. The simplest case would be a set of entirely myopic strategy choices. In such a context, agent populations pursue their search for higher fitness configurations assuming the other populations do not change the states of the C interdependent elements.

A second aspect relates to the payoff structure of the game itself. The likelihood of encountering stable strategies for elements exhibiting C externalities and thereby stability in co-evolution would be enhanced if the game was structured in such a manner that optimal choices given the behaviour of other agents coincide between groups. This refers to the existence of pure-strategy Nash Equilibria, i.e. the game has to be structured that one group of agents is best off by setting an element state to 0 or 1, regardless of the strategies chosen by other parties. Taking this aspect into account, however yields only part of the phenomenon: What has been highlighted so far relates to the existence of potentially evolutionary stable strategies with respect to strategy choices for the cross-group interdependencies (C). It partly relates to the stability of the entire co-evolving system: Deformations of agent fitness landscapes will cease as soon as a *Nash-Equilibrium* for the 'C-related' elements is found. This would allow each agent group to optimise the other n-C elements it controls independently. However, the existence of Nash Equilibria also relates to aggregate agent group behaviour which depends on the characteristics of the non-related n-C other elements controlled by agents in each group as well. It is by no means said that optimisation of the system's configuration would yield stable states for the C elements before reaching the (local) system optimum. The dynamics driving system optimisation therefore depend on the nature of the fitness landscapes for each co-evolving system population as well as the nature of search processes pursued.

As has been demonstrated more extensively elsewhere (Kauffman 1993, Chap. 6), the stability in co-evolutionary systems depends on their structure, i.e. the values for the control parameters N, K, C and S (number of interacting groups/ species). Simulating two reciprocally interacting populations (S=2) exploring their fitness landscapes through local search, it was found that the time before encountering stable solutions increased with C as large values of C imply strong interdependencies between populations. In such a setting, adaptive moves towards higher fitness by one group of agents distort the fitness landscape of the other group more strongly, i.e. they *spread* further within the system. This in turn would provoke adaptive moves in the second group with a reciprocal influence on the fitness landscape of the first. The more fitness landscapes are linked (the higher C), the more they deform if partners change their configuration implying a smaller likelihood of encountering stable configurations for both populations.[95] The value of K in turn influences the number of (local) optima in the fitness landscape of each population. High K values increase the ruggedness of landscapes implying that locally optimal configurations are encountered more quickly. This in turn stabilises the values of the C interdependent elements for both groups and thereby enables stability at the system level.[96] Average fitness of the co-evolving pairs is in turn high if C and K are either high or low. This relates to the issue of stability and interdependency described above.

Simulations conducted with different numbers of species (S) interlinked at C elements showed that the waiting time to encounter stability increased with S while the mean fitness of co-evolving partners decreased.[97] The explanation for this phenomenon touches upon the issues of the number of local optima and stability in co-evolving systems already outlined above. Allowing for an increase in the elements mutated in each search step did not improve the average fitness of co-evolving systems. Since a greater scope of search reduces the number of local optima in each group's fitness landscape, encountering stable system configurations becomes less likely. This leads to greater changes between the fitness landscapes of both groups due to their interdependence (C) and reduces their average fitness. Increases in overall fitness due to greater search scope are only possible if the number of 'long-jump' search steps is limited: A few of these more long-jump search steps might bring a population into the basin of attraction of another, potentially better (local) optimum (Kauffman 1993, p. 250).[98]

---

[95] The time before encountering Nash Equilibria was short when C<K and long when C>K (C=K approximately separated the two regimes).

[96] Due to a similar notion, simulations with fixed C and K but increasing N values showed that the waiting time for Nash Equilibria increased with N as larger values of N lengthen the time for encountering an optimum (especially with the local search activity assumed in these simulations).

[97] More precisely, Nash Equilibria are encountered quickly if K>SxC and very slowly if K<SxC (K=SxC acts as the separator between both regimes).

[98] While Kauffman's original work also includes the role of emergence and selection mechanisms for each subsystem, the results obtained are omitted here as both mechanisms are not included in the short-term model of clusters adaptation to change (see Chap. 4).

6.4 The behaviour of N/K systems: Agent, group and aggregate dynamics

The previous sections have highlighted that the dynamics of co-evolving NKC systems are shaped by agent and group dynamics as well as their interrelation at the system-level. The value of C externalities determines the extent to which the activities of different agent groups feed back onto each other. With high levels of cross-group interdependence, even small modifications in strategy within each agent group have significant repercussions on others and thereby the fitness of the system as a whole. High C values therefore tend to have a destabilising effect on co-evolution processes (all else held equal). Even with more limited degrees of cross-group interdependency, the stability of co-evolving systems depends on the likelihood of finding stable configurations for the elements linked by C externalities. This is all the more likely, the fewer changes occur within each agent group. The number of possible changes in turn relates to agent and group dynamics. At the group level, the number of possible improvements has been shown (6.2) to depend on the number of agents per group as well as the scope of the group's selection mechanism (geared at improving group or system fitness). The number of possible changes at the agent level in turn regards the nature of the fitness landscape of every agent (n, K) as well as the search and selection mechanisms employed.

To conclude, the stability and performance of adaptation in co-evolving systems depends on three aspects: The number of possible improvements in each subsystem, their optimality from a system perspective[99] as well as the spread of changes in the system, which is a direct consequence of the extent of interdependence between agent groups. On average, stability can be expected to increase with increasing K values and low numbers of elements modified per search step at the agent level. This leads to agent's lock-in with local optima, thus limiting the number of possible modifications. At the group level, greater numbers of agents and a selection mechanism geared at improving system fitness would mean that a number of possible configurations gets rejected throughout experimentation in the group and therefore do not affect the C related elements. Both aspects contribute to stability in system dynamics. Finally, low values of C help limit the spread of agent's modifications across agent groups. Co-evolving systems characterised by all these characteristics would be expected to exhibit more stable behaviour. The trade-off between the speed of finding better configurations and the performance and stability of system behaviour mentioned before (Sect. 6.1) thus also materialises in co-evolving systems. Chap. 7 will investigate how division of labour and co-ordination can act as a mediating force in this trade-off, thereby determining the adaptability of different cluster types.

---

[99] Related to the selection mechanism for modifications at the level of agents in each subsystem (level of fitness to be improved), the selection mechanism employed in a group (improving group or system fitness), as well as the number of agents per group.

7 Clusters as co-evolving N/K systems

As has been spelled out in the previous sections, clusters can be understood as complex, co-evolving systems, where different agent groups control a subset of the system. Agent activities are influenced by the rules provided in the local culture and their outcome exhibits an interdependence spanning across groups, which is attributable to the effects of agglomeration externalities. It has also emerged from Chaps. 5 and 6 that analysing the behaviour of such systems from an N/K perspective requires two things: A definition and setting of values for the different control parameters (N, K, C and S) as well as an understanding of the dynamics of and interrelation between the different levels of the system (agent, group, system).

This section therefore starts by sketching out how clusters can be depicted in the N/K framework. More precisely, Sect. 7.1 outlines what elements of the cluster constitute the control parameters N, K, C and S and how agent, group and system dynamics relate to one another. From these aspects, part 7.2 develops a rationale for including the division of labour and the mode of co-ordination in clusters into the model and shows how both factors can steer the self-organisation process of agent-driven adaptation towards more or less successful results. The division of labour and mode of co-ordination will be accommodated in the framework by increasing interdependence between agent activities on the one and altering the goals underlying agent activities on the other hand. The final section in this chapter then outlines the intuition to be tested with the simulation model. A presentation and discussion of findings will follow in Chap. 8.

7.1 Parameters, agent, group and system dynamics

According to the understanding of clusters provided in Sect. 4.1, the latter consist of agents conducting activities in the value chain of one or several related industries. Their activities are interrelated due to the existence of agglomeration externalities and co-ordinated by the orientation provided in a local culture. In other words, clusters consist of a set of activities in an industry directed at the manufacture of a set of products or services (the local value chain). These activities are performed and thus controlled by different local agents (firms, research and training or policy units). Due to this decentralisation (non-integration) of production activity, clusters exhibit a horizontal dimension with competing agents conducting similar activities and a vertical one where agents' actions are complementary. To simplify analysis, it will be assumed that all agents in the horizontal dimension of the cluster control the same set of activities and that there is no overlap in activities between agents at the vertical level. All end-producers will thus perform the same range of activities and the activities of end-producers will differ from those conducted by suppliers (including service providers), research and training or policy units. It will furthermore be assumed that industry activities can be differentiated sufficiently in order to avoid their multiple allocations to actors. For instance, 'research' could be represented by different elements (market and basic research, testing etc.) to allow for an allocation of each task to one actor group (e.g. firms, research and training units or service providers).

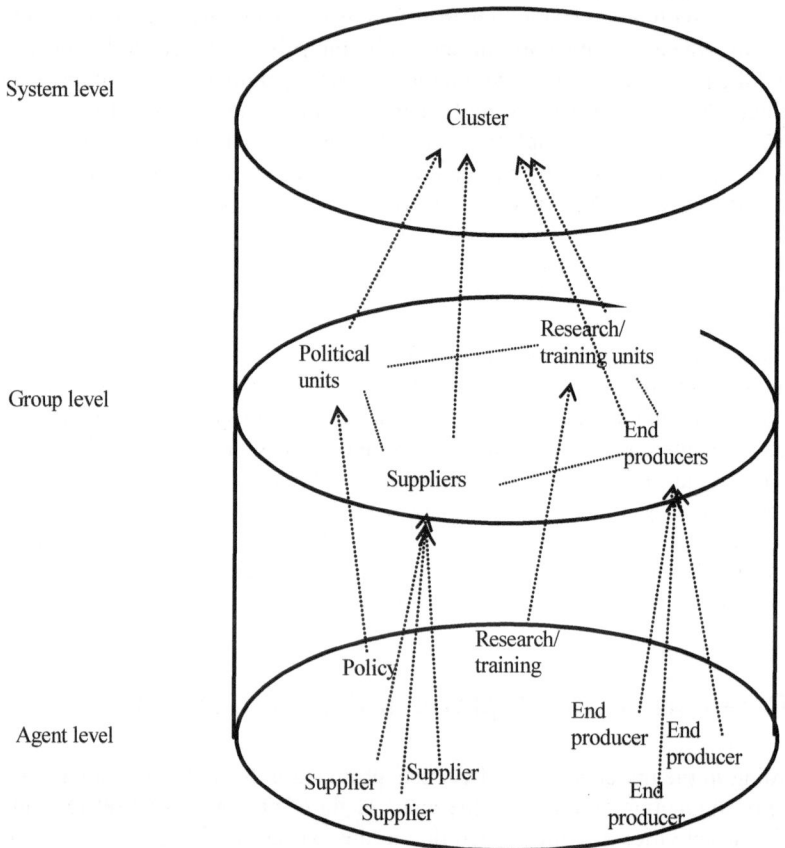

Fig. 7.1. Agent, group and system dynamics in clusters

At the lowest level of analysis (Fig. 7.1.), individual agents conduct a specific range of activities. Grouping them according to the similarities in their activities gives the horizontal dimension of the cluster (group level). Determining the aggregate group from individual agent behaviour, and relating it to the outcome at the system-level involves moving from the horizontal to the vertical dimension of the cluster, where the activities of agent groups are interdependent. This interdependence stems from the effects of agglomeration externalities: The existence of positive and negative externalities from proximity means that the activities of one group of agents can have a positive or negative feedback on the outcomes of activities conducted in other groups. Agglomeration economies determine how agent groups can benefit from certain activities of others while negative externalities reduce the success of group actions. Compared to agent groups acting in isolation, these situations therefore imply that the joint activities of groups linked by agglomeration economies outperform those of both groups in isolation. In the case of negative externalities, the performance of certain activity combinations is lower as compared to their execution in isolation.

Taking agent and group behaviour as well as inter-group interdependencies into account yields the system's performance by showing the current configuration of the cluster and its fitness in a specific environment. To understand the key factors for cluster adaptability, one has to provide an understanding of the dynamics and interdependencies of the three different system levels through accounts of:

- the behaviour of agents,
- the relationship between individual and agent group dynamics in the horizontal dimension of the cluster and
- the effects of inter-group interdependencies at the vertical dimension of the cluster.

As was found in Chap. 6, each of these aspects is determined by the control parameters of the system (N, K, C and S), which is why the following section first outlines what constitutes the latter in the context of clusters.

7.1.1 Model parameters: N, K, C and S

The N/K model proceeds by describing any co-evolving complex system by the number of its elements (N), the intra- and inter-agent interdependencies (K, C) as well as the number of agent groups (S) in the system. The most elementary units of analysis are therefore those elements that are controlled by different agents and that can be linked within and between agents. In the case of clusters, N corresponds to the different activities conducted in the local value chain. Agents (firms and other organisations) control these activities, which can be interdependent at the level of individual actors as well as between agents.

Interdependencies in the N/K framework then imply that the success of one activity depends on the strategy (choice of an element's state) adopted with respect to another. At the level of firms (K) this reflects that strategy choices with respect to different activities (e.g. research and production) have to be aligned to produce good results, i.e. not all possible strategy combinations perform equally well. Interdependencies between the activities of different agents then mirror the effects of agglomeration externalities C. In the presence of knowledge spillovers, for instance, research undertaken by one firm can be more successful if another actor invests in research activity as well. At the same time, the existence of agglomeration externalities also implies that certain strategy combinations between agents have a negative impact on their performance. In the case of activities linked by agglomeration economies, reductions in the individual effort by one agent can negatively impact on the success of activities undertaken by another (by taking away the 'spillover' effect conveyed by positive externalities). Negative agglomeration effects in turn imply that specific strategy combinations adversely affect the agents linked by the externality as it transmits the negative impact of one or several strategies undertaken (e.g. pollution effects). Positive feedbacks between activities in the N/K model thus indicate that thanks to an interdependence between say n_1 and n_2, the joint fitness contribution for a specific combination of element states $w(n_1=0 \cap n_2=0)$ exceeds the sum of fitness contributions for those two very element states in independent systems, i.e. $w(n_1=0 \cap n_2=0) > w(n_1=0) + w(n_2=0)$. The opposite holds for negative feedback effects. As a result, the effects of agglomeration externalities are captured by the parameter C. This understanding of agglomeration externalities is intentionally broad, encompassing both positive and negative feedback effects between firm activities in co-location. It extends beyond the traditional Marshallian agglomeration economies to include diseconomies from agglomeration as well.

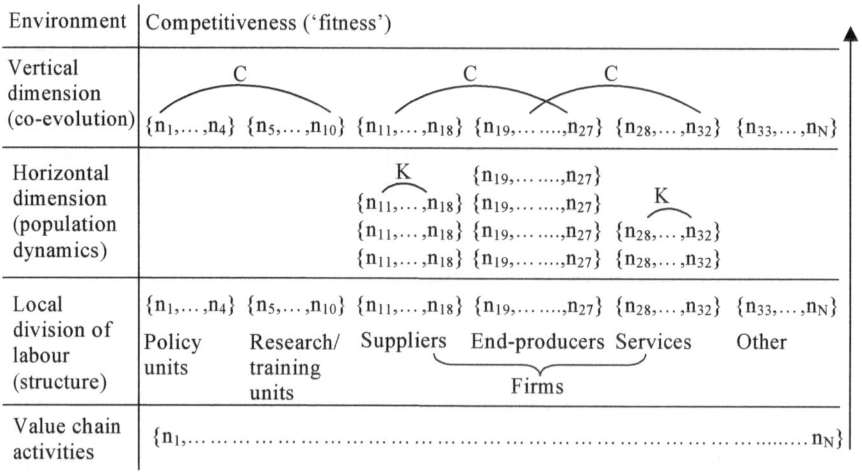

Environment	Competitiveness ('fitness')
Vertical dimension (co-evolution)	C over $\{n_1,...,n_4\}$ $\{n_5,...,n_{10}\}$; C over $\{n_{11},...,n_{18}\}$ $\{n_{19},.......,n_{27}\}$; C over $\{n_{28},...,n_{32}\}$ $\{n_{33},...,n_N\}$
Horizontal dimension (population dynamics)	K over $\{n_{19},.......,n_{27}\}$ $\{n_{11},...,n_{18}\}$ $\{n_{19},.......,n_{27}\}$; K over $\{n_{28},...,n_{32}\}$ $\{n_{11},...,n_{18}\}$ $\{n_{19},.......,n_{27}\}$ $\{n_{28},...,n_{32}\}$ $\{n_{11},...,n_{18}\}$ $\{n_{19},.......,n_{27}\}$ $\{n_{28},...,n_{32}\}$
Local division of labour (structure)	$\{n_1,...,n_4\}$ $\{n_5,...,n_{10}\}$ $\{n_{11},...,n_{18}\}$ $\{n_{19},.......,n_{27}\}$ $\{n_{28},...,n_{32}\}$ $\{n_{33},...,n_N\}$ Policy units / Research/training units / Suppliers / End-producers / Services / Other Firms (Suppliers, End-producers, Services)
Value chain activities	$\{n_1,...n_N\}$

Fig. 7.2. Clusters as co-evolving systems – Applying the N/K framework

The total of N value chain activities is then allocated to different agent groups, i.e. there is a certain degree of division of labour at the *vertical* dimension of the cluster. In the N/K model, this is mirrored by the fact that the entire string of N elements is decomposed into smaller substrings of n elements which are controlled by different agent groups.[100] As a consequence, the horizontal level of the cluster is represented by a group of agents that acts upon a subset (or substring) of n elements (activities) of the system: A certain range $\{n_1,...,n_4\}$ of value chain activi-

---

[100] The number of substrings multiplied with the number of elements per substring has to yield the number of elements in the entire string, i.e. N.

ties is conducted by policy units while another range $\{n_{19},\ldots,n_{27}\}$ pertains to end-producers and so on (see also Fig. 7.2.). The smaller the substrings are, the greater the extent of division of labour (see also Sect. 7.2.1). The different agent groups controlling interrelated substrings of the entire system correspond to the interacting 'species' S in the N/K model. The greater the division of labour at the vertical level of the cluster, the smaller the range of activities controlled by individual agents or agent groups (n) and the greater the number of different agent groups (S) – provided the total number of activities in the local value chain (N) is held equal.

To sum up, when applying the N/K model to the case of clusters, the following definitions for the main control parameters are derived. N denotes the number of industry activities in the local value chain directed at producing a (set of) marketable product(s) or service(s). A subset of these activities n is controlled by each group of local agents. S in turn measures the number of agent groups controlling different substrings (activity sets) in the cluster. This parameter reflects the degree of local division of labour at the vertical level of the cluster. Activities can be interdependent at the level of individual agents (K) and between agents (C). The latter aspect reflects the effect of agglomeration externalities inducing interdependence of activities in the local value chain. In its current conceptualisation, the model only accounts for agglomeration externalities at the vertical dimension of the cluster. However, externalities especially those regarding learning are likely to be found at the horizontal level as well. To account for this, a bidding process (see Sect. 7.1.4) is introduced reflecting strong learning from observation at the horizontal level of the cluster by mirroring a strong diffusion of best practice within agent groups.

7.1.2 Model setup: Agents, elements and interdependence

All simulations were run using the LSD (Laboratory of Simulation Development) platform[101] for clusters containing N=24 elements with increasing levels of fitness landscape interdependency, i.e. increasing values of the corresponding LSD parameter (EvenK). The value of this parameter has important implications since the programme departs from the original N/K formulation by allowing for both a random as well as a regular distribution of element interdependencies. In the context of the latter, the programme parameter EvenK denotes the length of a 'block' of interdependent elements. For the modelling of cluster adaptation to change, this implementation of regularly distributed interdependencies was adopted for two reasons.

First, it allows for an easier determination of the extent of intra- and inter-agent interdependencies. By taking the size of interrelated element blocks as well as the

---

[101] LSD is a freely available shareware programme covering a wide range of economic simulation models. I am heavily indebted to Marco Valente for his guidance and support in developing a workable simulation model. Interested users can find LSD online: http://www.business.aau.dk/~mv/Lsd/lsd.html. All code and simulation data for the model presented here are available from the author.

degree of division of labour in the model clusters, it is possible to simply count element interdependencies at the level of agents (within a substring) as well as between agents (between substrings). Dividing these absolute numbers by the total number of elements (N) will yield the average interdependence for each system element with other elements at the level of the agent (K) and between agents (C). Since both aspects matter for the adaptability of co-evolving systems (= their ability to find better configurations), such a systematic approach is preferable to generating a landscape with random interdependencies.[102] Second, the choice of fitness landscapes with block interdependencies can be justified by the topic of this study. In the case of cluster adaptation to change, the distribution of inter-agent externalities matters (see also Chap. 5) by affecting the extent and range of interdependencies between them. This determines which agents are affected how strongly by the activities of others. In this context, it can be argued that activities (elements) that are closer in the overall local value chain are more interdependent as they can be expected to benefit more from the agglomeration externalities represented by C interdependence (see Sect. 7.2.1).

To avoid effects of one-off 'lucky' adaptations, 100 simulations were conducted for each EvenK parameter value. When investigating the effects of environmental bifurcations, these simulations were run starting with a new fitness landscape every time corresponding to a cluster adapting to 100 bifurcation events that altered its entire competitive environment. In the case of external perturbations, the fitness contributions of a limited number of elements (six out of 24) were changed. To obtain comparable results to the case of bifurcations, this shift in fitness contributions occurred 100 times during the simulation. Adaptation was allowed for 600 simulation steps following each bifurcation or perturbation event. Throughout this time, agents were allowed to change and evaluate their configurations.[103] Average fitness values and standard deviations were gathered both at the system, i.e. cluster and at the (firm) group level.

Different types of clusters then compete for the best result in adaptation. They differ in their architecture regarding the extent of division of labour on the one, and the mode of co-ordination on the other hand. The respective cluster architectures are argued to have evolved throughout their development. Within the more short-term perspective of agents self-organising their immediate adaptation to change adopted here, such architectures are *given*. Agents cannot alter the range of activities under their control or the local rules of the game in the short term, but have to adapt with their actions being constrained by both aspects (see also Sect. 7.2). Comparing the adaptive performance of clusters with different architectures then yields an explanation for the roles of division of labour and co-ordination for cluster adaptation within an N/K framework.

---

[102] In this approach one would have had to investigate which parameter values this random distribution yielded for every simulation run, implying a considerable risk of omission or repetition for different parameter values.

[103] In the case of bifurcations, 100 simulations with 600 steps each were conducted. In the case of perturbations, one simulation with 60,000 steps and a shift in the landscape occurring every 600 steps was run.

System decomposition (setup)

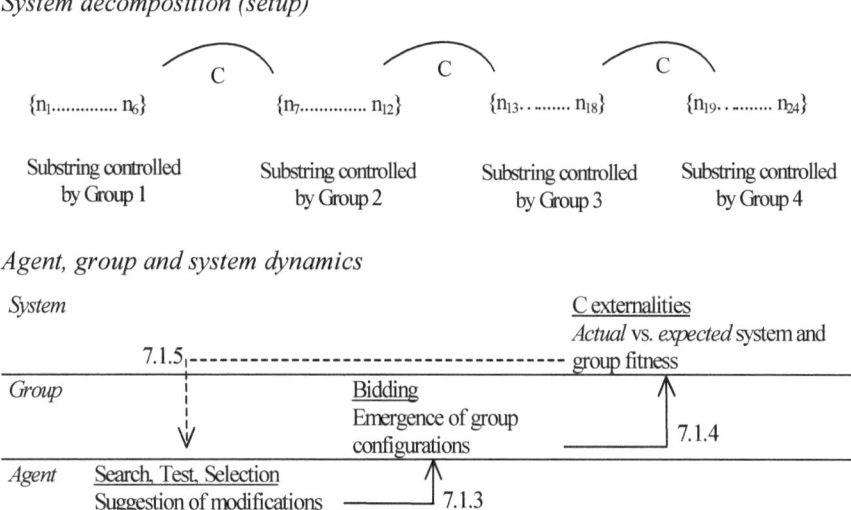

Agent, group and system dynamics

Fig. 7.3. The model in schematic representation

The different clusters all exhibit the same division of labour with S=4 substrings containing n=6 out of 24 elements (Fig. 7.3.).[104] This division of labour was adopted to allow for a sufficiently large search space at the level of substrings (2^6 possible configurations with a minimum of 64 different fitness values in independent fitness landscapes). Finer divisions of labour might have influenced the simulation results by narrowing the search space to an extent that search outcomes alone (rather than division of labour or mode of co-ordination) would have driven simulation results. A sufficiently large search space was thus needed to allow for enough possible system and substring configurations so that different modes of inter-agent co-ordination as a selection mechanism will matter even with low degrees of fitness landscape interdependence (reflecting a limited division of labour; Sect. 7.2.1).

Each of the substrings is controlled by one group of agents. The agent group consists of five agents that all conduct the same range of activities (n) within the substring. Each agent controls his own substring composed of six elements. In total, all clusters consist of 20 agents that are evenly distributed between four groups. This setup was selected to single out the role of different co-ordination mechanisms. Since the number of agents per group matters for the adaptability of the cluster (see also Chaps. 6 and 8), all cluster groups in the simulation model contain the same number of actors. For details on the computational implementation of the simulation model, especially with respect to the following agent, group and system dynamics see appendix 5.

---

[104] The role of different divisions of labour was indirectly accounted for by increasing the EvenK parameter (see also Sect. 7.2.1).

7.1.3 Search, test, selection: Agent dynamics

All agents search the fitness landscape of their substring by altering the state of the elements under their control with a probability of 0.5, i.e. on average, they change three out of six elements. This intermediate search process was chosen to give agents the chance to discover the global optimum in their subset of the fitness landscape while not relying exclusively on the unlikely event of a successful long-jump. Agents then evaluate the fitness and (possibly) select the new configuration of their 6 elements according to different criteria representing the mode of inter-agent co-ordination in the cluster (see Fig. 7.3.). The evaluation of the new configuration is done according to its *expected* fitness, i.e. the fitness obtained while holding the remaining 18 elements outside the agent's control constant. In a sense, agent search and selection activity is thereby *myopic*. Two aspects about clusters speak in favour of this perspective.

The first regards the notion of bounded rationality of cluster agents with respect to the exact extent and the mechanisms underlying agglomeration externalities. In other words, firms might not know that their good innovative performance partly relies on knowledge spillovers provided by competitors. Second, even if actors do know that they rely on the activities of others, it is highly unlikely that any individual agent can determine the exact extent of his or her individual contribution to agglomeration externalities. For instance, an individual firm cannot foresee how an increase or decrease in its labour training activity will affect the pooled labour market in the cluster. This uncertainty is due to the fact that the impact of agent activities on the nature and extent of agglomeration externalities also depends on the strategies chosen by others. It would be impossible for any individual actor to be fully informed about the plans and future strategy choices of all other agents. Choosing a strategy that will work well in the current context is therefore any actor's best bet. This implies that individual activities may miss their goals or lead to unanticipated aggregate effects.

The difference between the clusters modelled here then resides with their selection mechanism, i.e. with the fitness level that has to be improved for a modification to be chosen by individual agents. The selection process at the agent level mirrors different co-ordination mechanisms in clusters that are argued to impact on agent strategy choice. This is thus not to be misunderstood as a form of 'collusion' of agents where they jointly agree on the best practice possible given specific goals. Instead, co-ordination is achieved by establishing a certain mindset within individual agents according to which they select their strategies. Again, the different modes of co-ordination described below are taken as given constructs and their role for steering agent adaptation towards good collective outcomes is investigated.

The emergence of any specific mode of co-ordination lies with the intuition developed on the local culture (see Sect. 4.1): To avoid a tragedy of the commons with respect to the activities underlying agglomeration economies, local agents agree on a set of rules of the game to be followed by all cluster actors. Defection from these rules is punished by use of formal institutions (whenever applicable) or by collective action excluding a defecting agent from access to all other cluster networks conveying positive agglomeration externalities (Cappellin 2003). As a result, defection in one business context can entail costs well beyond that very context itself.[105]

The existence of local rules of the game then produces an incentive for all agents to adhere to them, i.e. to gear strategies at compliance with the local culture. Depending on the nature of local rules, this leads to very different agent goals. As a consequence, the existence of rules on acceptable business practice *indirectly* leads to a co-ordination of agent activities by influencing the goals underlying strategy choice. One can therefore speak of the existence of an indirect mode of co-ordination in clusters that is attributable to the emergence of a set of local rules on acceptable business behaviour which is often dubbed the 'local culture'. Regarding the relationship between agent goals, strategies and local rules of the game, a reciprocal influence (Sect. 4.1) emerges: Through their goals and strategies, agents in the cluster create and alter the rules in the long run. At the same time, adherence to the local rules shapes the goals underlying agent strategies. In the short term of self-organised adaptation to change, local rules thus constitute fixed characteristics limiting agent behaviour by determining feasible strategies.

With respect to the modes of co-ordination provided by local rules of the game, four ideal-typical instances are modelled (for further detail see Sect. 7.2.2):

- Even distributions of power: Individualistic and collective clusters
  Both instances represent clusters with an even distribution of power. In the first case (*individualistic*), agents care only about their own fitness when selecting a strategy. Adaptation in the second case (*collective*) is instead geared at improving cluster fitness as a whole.
- Uneven distributions of power: Alliance and leading firm clusters
  In alliance clusters, two agent groups have formed *alliances* to improve their joint fitness. In the leading firm case, one firm has come to *dominate* the cluster, i.e. all other agents seek to improve their fitness and that of the dominant firm while the latter behaves egoistically, seeking to maximise its individual fitness.[106]

Since agents simultaneously decide on a strategy and the effects of inter-agent interdependencies (agglomeration externalities) are unknown to them beforehand, it is entirely possible that an agent's goal remains unfulfilled – either because her activities turn out to be harmful to her aims thanks to adverse effects on the activities of other agents, or because a specific activity is only successful in the presence of a certain strategy of another actor. These events will be more or less likely depending on the extent of division of labour and the mode of co-ordination underlying agent strategy choice (see Sect. 7.2).

---

[105] It is this very notion of local rules of the game that underlies many of the observable integration problems faced by multinational corporations when locating into an existing cluster (see Heiduk and Pohl 2002 on the case of European MNEs in Silicon Valley).

[106] In the leader case, the group containing the dominant firm has only one agent, i.e. there are 16 agents in total.

7.1.4 Bidding for representation: From agent to group behaviour

To go from individual agents to a new configuration at the substring (group) level, a bidding process is introduced for each agent group (Fig. 7.3.). In it, agents propose their respective configurations (obtained through search and selection processes) to be chosen as the new configuration for the entire substring. In other words, agents bid to be the representative for their group at the level of the cluster. For each group, a 'champion' is selected holding the other group configurations constant. Similarly to the case of individual agents, the *expected* fitness of a new configuration is therefore evaluated in each group, assuming that agents in other groups do not change their activities. The agent with the best configuration regarding expected cluster fitness is then chosen to be the new substring configuration for the next simulation step. This could be argued to represent the learning from observation that is prominent at the horizontal level of clusters, i.e. within each agent group (Maskell 2001): A strong learning from observation would lead to a selection and imitation of best practice. In the case of clusters where the success of firm activities depends on those of others, such best practice corresponds to strategies that work well in the context of other agent (group) activities. To approximate this aspect, the bidding process is based on the highest expected cluster fitness rather than individual agent or agent group fitness.

This group selection criterion (cluster fitness) was chosen for a number of reasons. First, in a setting where individual firms produce a product together with other agents, strategies that maximise any individual actor's fitness at the expense of the entire system would lead to inferior outcomes of system dynamics. In a cluster trying to produce a marketable product, such strategies would be selected against by market and competition forces. As an example, cost cutting by supplier firms at the expense of input quality would harm the quality of the final product and thereby the entire cluster as well as (in the long run) the firms that initiated this adverse change in strategy. Second, in interdependent systems where agents depend on one another, best practice comes in the guise of strategies that work well with other agent strategies. While this may not necessarily lead to an immediate maximisation of agent (group) fitness when the latter is viewed in isolation, the fact that strategies have repercussions on other cluster agent groups as well as the other way round makes adopting activities that work well within a given context of actions by others the most viable strategy. Given this nature of the bidding process, the number of agents in each group will be important for the number of different configurations tried at this level and thereby the expected fitness level attainable in a cluster. To reflect the relatively short-term nature of the problem under study here (adaptation after a local bifurcation), the number of agents in each group is kept constant over time, i.e. selection dynamics are not included.

7.1.5 Deforming landscape subsets: The dynamics of co-evolution

In the next step, the agent configurations that won the bidding process in each group are 'fed' into the entire fitness landscape. Taken together with the interdependencies due to agglomeration externalities (represented by C), this yields a new cluster configuration which in turn determines the *actual fitness* of the cluster and its agent groups (Fig. 7.3.). Based on this new configuration, the *expected* substring and cluster fitness can be compared to the *actual* fitness and might call for a revision of agent strategies (see also Allen 1997; Axelrod 1997 or Axelrod and Cohen 1999). Agents thus act myopically and – depending on the extent of C externalities as well as the actions of others – do or do not reach their goals in each time step. A key feature of co-evolving systems therefore lies with the fact that they are not about attaining a (local) optimum as was the case in single N/K systems. Instead, in co-evolutionary processes, "*the landscape of one actor heaves and deforms as the other actors make their own adaptive moves. [...] Thus co-evolving behaviour is in no way limited to attaining point attractors which are local optima, not is it even clear that coevolving systems must be optimizing anything whatsoever*" (Kauffman 1993, p. 238). In a sense, inter-agent interdependencies thus act as a disturbing factor on individual agent adaptation. Depending on the nature of this disturbance, agents may need to revise their strategies in subsequent simulation steps. This co-evolutionary process can settle into stability or remain chaotic depending on the structure of the underlying fitness landscape, especially the extent of cross-agent externalities (see also Chap. 8).

One might argue that the assumption of full reversibility of agent strategies underlying the model is unrealistic in the case of clusters where firm choices with respect to their activities could be costly to undo.[107] Including a cost of strategy reversion would be possible by arguing that agents require a threshold improvement in the expected fitness level upon which their selection mechanism is based before executing any modification. Since the model assumes homogeneous agent characteristics, such an implementation would only reduce the number of selected agent modifications and stabilise adaptation processes for all cluster architectures. The relative performance of different architectures lying at the heart of the analysis conducted here would be unchanged by this aspect. As a result, the model does not account for a cost of reversing decisions at the level of cluster agents. Differences observed in the adaptiveness of different cluster types with fully reversible strategies would surely materialise even more strongly in a context where strategies are (partly) irreversible.

7.2 Accommodating division of labour and mode of co-ordination

The division of labour and mode of co-ordination influence all three levels of cluster dynamics and hence its adaptation process. The extent of division of labour here regards the degree to which the local value chain is divided in the vertical dimension, i.e. the division of labour mirrors the level of vertical differentiation between cluster agents. With a growing division of labour (all else held equal),

---

[107] I would like to thank Peter Maskell for pointing out this limitation.

less value chain activities are conducted within an individual agent. This will be shown to determine the fitness landscape structure for both the substrings controlled by different agent groups as well as the extent to which activities are linked between them. A growing division of labour in clusters is thus associated with growing interdependence of activities both within and across groups. This decreases the likelihood of finding both good and stable solutions in adaptation processes. The mode of co-ordination in a cluster is argued to influence the goals underlying agent strategy selection. This will be mirrored in different selection mechanisms, which determine the likelihood of finding superior configurations in each simulation step. This likelihood decreases with more collective orientations, leading to both better adjustment results from a system perspective and a greater stability of adaptation processes.

7.2.1 Fitness landscape structure: The role of division of labour

In the simulation programme used (LSD), the parameter EvenK denotes the size of element blocks that are mutually interdependent. Such a conceptualisation may be more realistic regarding the notion of agglomeration externalities in clusters. It is reasonable to ascertain that activities that are closer to one another in the value chain will be more affected by agglomeration externalities. For instance, suppliers and end-producers are likely to be more positively influenced by their respective activities than suppliers and some general service providing organisations. As a result, rather than assuming randomly distributed element interdependencies, clusters will be understood as potentially decomposable systems (Sect. 6.1) where element linkages are distributed into blocks. As a result, EvenK denotes the size of such element blocks. For example, if EvenK=4, elements {1-4}, {5-8}, {9-12}, {13-16}, {17-20} and {21-24} are linked to one another and can be grouped into 'blocks' of the size four. With a given decomposition of the system cluster into agent groups controlling a substring of elements, increasing the size of element blocks (higher EvenK values), leads to higher interdependencies within and between cluster agents, i.e. higher K and C values.

The extent of division of labour has already been found to be a question of decomposition of the entire value chain at the vertical dimension into substrings controlled by different agent groups. The finer the division of labour, the smaller the range of activities conducted by agents in each agent group. As a result, for a given length of the value chain N, a greater division of labour means that more agent groups (S) are controlling a smaller subset of activities (n). For a given number of activities in the local value chain (here N=24), and a given extent of interdependence (here EvenK=8), a greater division of labour corresponds to greater values of cross-agent externalities. For low EvenK values, a greater division of labour also corresponds to higher K values. However, the values of K are limited by the number of elements in the substring n. As a result, $K \leq n-1$, i.e. once the maximum K value is reached, it cannot increase, no matter how much EvenK grows. One can therefore not say that a greater division of labour is always associated with higher K values (see also Fig. 7.4.).

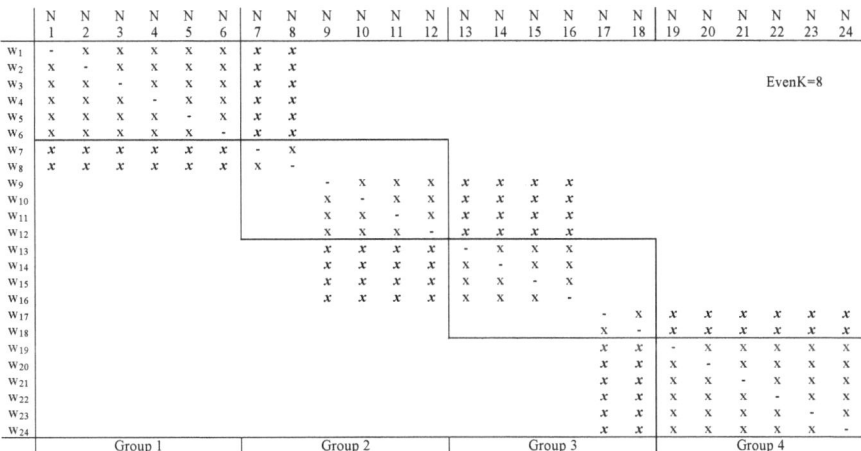

Number of intra-agent externalities = 88, i.e. K=3.67[108]
Number of inter-agent externalities = 80, i.e. C=3.33

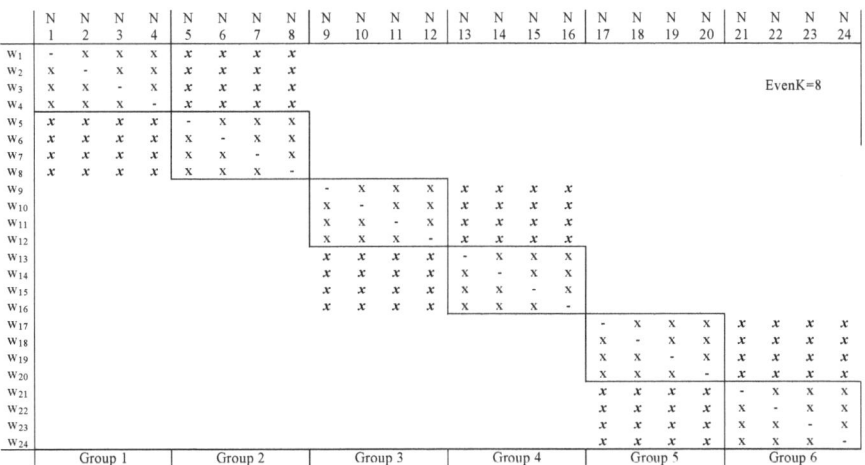

Number of intra-agent externalities = 72, i.e. K=3
Number of cross-group externalities = 96, i.e. C=4

Fig. 7.4. Division of labour and N/K(C) parameters

The complexity of systems in the case of clusters (K, C) is therefore a result of the degree of division of labour in them. The smaller the range of activities conducted

---

[108] The value of K (average number of within-agent elements influencing an element) was derived by dividing the number of intra-agent externalities (88-all black x) with the total number of elements (24). This yields K=3.67. Similarly, the average number of inter-agent interdependencies was derived from the total number of C externalities (80, **bold x**) divided by the number of elements (24), i.e. C=3.33. The K and C values for all other EvenK fitness landscapes were determined accordingly (see also the fitness landscapes in appendix 4 and the results overview in appendix 6).

by individual firms, the greater the number of C externalities for any given degree of element interdependence. In a way, a stronger decentralisation would therefore be associated with higher C values. This idea may run counter the notion of modularity as a precondition for a division of labour between actors.[109] In the outsourcing debate, a certain degree of modularity acts as a precondition for a division of labour between independent firms. This would imply that very interrelated activities remain within the control of one integrated firm (e.g. Ethiraj and Levinthal 2004b; Langlois 2002 or Siggelkow and Levinthal 2003). As has been suggested elsewhere,[110] this notion does not necessarily apply to the case of clusters. Due to their superior means for observation and interaction between co-located agents, clusters provide a form of market organisation that is more integrated than the pure market transactions prominent in the outsourcing debate, yet less integrated than common ownership (Lorenzen and Maskell 2004). As a result, clusters allow for a division of labour even in the presence of interdependencies between activities (Helsley and Strange 2003). This is also in line with the observation that clusters develop a deeper division of labour than would be possible between distant firms (Malmberg and Maskell 2002; Maskell 2001).

The extent of division of labour and cluster adaptability will be shown to exhibit a trade-off. When comparing adaptive performance in clusters with different degrees of division of labour (decentralisation), very decentralised clusters (high division of labour) will perform worse than more integrated ones (all else held equal).[111] This is due to the overwhelming problems associated with optimising a system characterised by very high C externalities. At intermediate stages (medium C values), more decentralised clusters might however outperform more integrated ones: The smaller range of elements controlled by an agent (n) in decentralised clusters is faster and easier to optimise than the greater number associated with large, integrated actors (see also Sect. 6.1.2).

One way to test this intuition on the role of the division of labour is to compare the adaptive performance of clusters with different decompositions for a given EvenK parameter (e.g. clusters with N=24 elements that are decomposed into substrings of size n= 4, 6 or 12). However, a second possible way of investigating the role of division of labour exists. In an N/K framework the impact of division of labour on the dynamics of clusters is primarily driven by its structural influence, i.e. the values of K and C in each case. The notion that a growing division of labour increases inter-agent interdependencies therefore allows for an indirect treatment of the question by running simulations for increasing C interdependencies with an identical number of elements per agent (i.e. identical system decom-

---

[109] I am grateful to Koen Frenken for highlighting this issue.

[110] There have been suggestions to understand clusters as a co-ordination mechanism (Maskell and Lorenzen 2003) exhibiting transaction costs between those of integrated firms and market transactions (see Coase 1937 or Williamson 1975). Others (Helsley and Strange 2003) have viewed clustering as a substitute for integration. This implies that it is possible to outsource activities between co-located firms that would otherwise remain inside a firm.

[111] As long as the co-ordination mechanism and the number of agents per group are identical, the following notions hold. With multiple differences between clusters, tradeoffs between different determinants emerge (see Sect. 9.1.2).

positions). In a way, less interdependent systems (low C values) then give an idea of the dynamics of more integrated clusters while increasing fitness landscape complexity points at greater decentralisation. Put differently, by increasing EvenK, the effects of a greater extent of division of labour are indirectly accounted for in the sense that simulations for larger EvenK parameter values represent cases of clusters with a finer division of labour (see Fig. 7.5.).

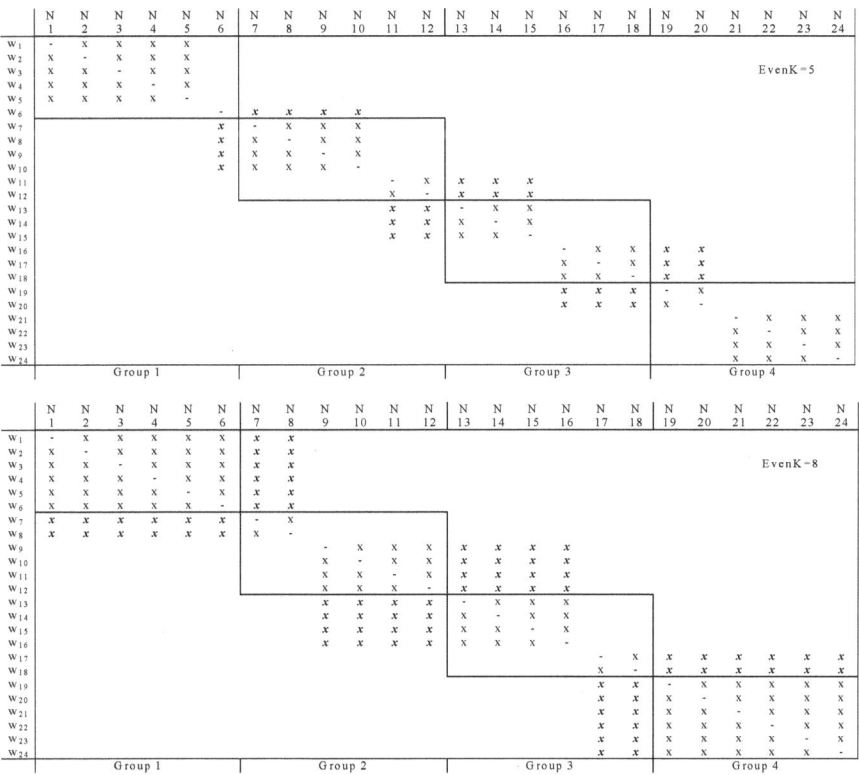

Fig. 7.5. Selected values of 'EvenK' and corresponding fitness landscapes[112]

---

[112] All fitness landscapes are given in appendix 4.

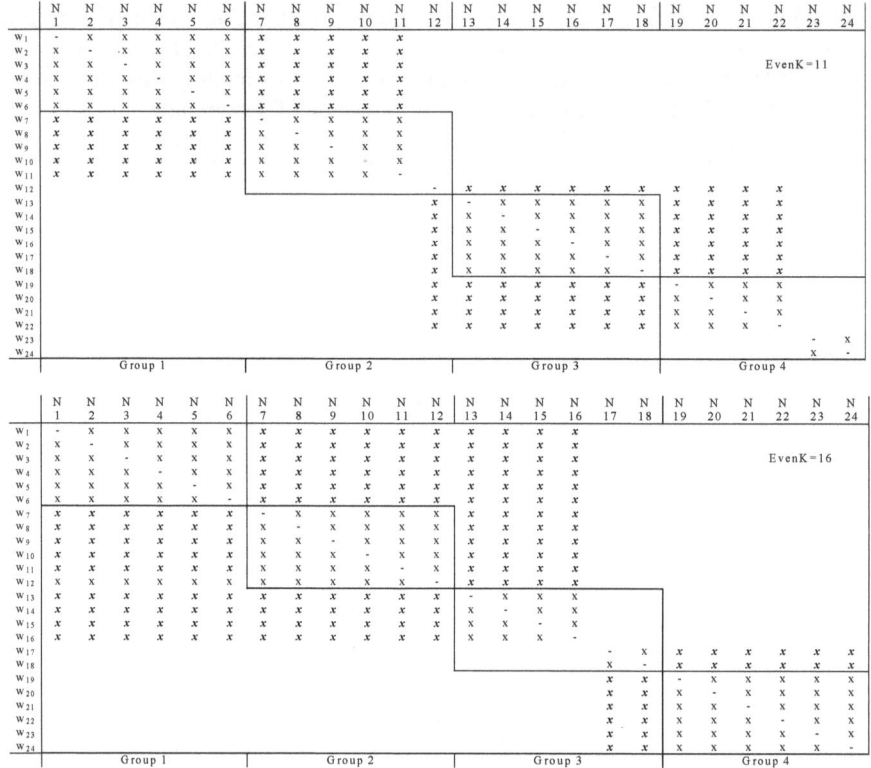

Fig. 7.5. (Cont.)

The extent of division of labour (indirectly measured through the EvenK parameter) then impacts on the behaviour of the cluster and its agents in two ways. First, the parameter determines the values of K and C for a given decomposition of the N=24 cluster elements (with n=6 and S=4). As a result, the division of labour feeds back onto the fitness landscape structure of individual agents as well as on the dynamics of co-evolution. As EvenK increases, so do the number of intra- and inter-agent externalities. Increases in K then imply that the fitness landscape for individual substrings controlled by the different cluster agents becomes more and more 'rugged'. As a consequence, there are increasing numbers of local optima allowing for agent lock-in. This aspect has a stabilising effect on agent behaviour implying that agents make fewer changes in their configuration as K increases.

At the same time, increases in EvenK lead to an even more pronounced increase of cross-agent externalities C. This has a destabilising effect on the behaviour of clusters as even small changes in an agent's configuration have large repercussions on the fitness of all other agents. Different values of EvenK, K and C will therefore lead to different cluster dynamics. Second, due to the way the division of labour is accounted for, it influences the relative centrality of different agents. Centrality is here to be understood as in social network theory, i.e. central cluster agents (groups) generate and receive a large share of C externalities. As EvenK increases, interdependence between groups one and two, as well as three and four grows. Once EvenK>12, agents in groups one and two become more central to the cluster, whereas group four is increasingly isolated (until EvenK>18). This aspect will play a role for the optimality of different modes of co-ordination in the simulation (see also Sect. 7.2.2).

To sum up, the extent of division of labour will matter for cluster behaviour in two ways. First, a growing division of labour increases interdependence between cluster groups implying a greater difficulty to optimise their behaviour with respect to the C-interdependent elements (Chap. 6.3). Second, a strong division of labour leads to differing centrality of groups. This in turn impacts on the effectiveness of co-ordination mechanisms. As will be elaborated further in the following section, co-ordination mechanisms are meant to internalise C externalities into each agent's decision making. Modes involving more central cluster agents are thus likely to outperform those involving less central ones.

7.2.2 Strategy selection: The role of inter-agent co-ordination

The co-ordination mechanism in the cluster affects agents' goals, i.e. the criteria that have to be met for an agent to select a modification discovered during search activity. In the N/K model, fitness acts as the selection criterion: Modifications of element states are only retained if they improve fitness. However, in systems with multiple agents engaging in decentralised search activity over the subset of the system they control, the level of fitness to be improved by search strategies can differ. For instance, agents could gear search activity within their decomposition at the improvement of overall system fitness. As a result, element modifications within the agent's scope of control would only be accepted if they improved average system fitness. The model introduced here takes four different selection criteria employed by all agents in the cluster as a proxy for the underlying modes of co-ordination:

- *Individualistic*: Agents seek to maximise their own fitness, i.e. that of the elements under their control,
- *Collectivist*: Agents want to maximise the fitness of the entire system,
- *Alliance*: Agents seek to maximise their own and partner fitness,
- *Leader firm*: Agents aim at maximising their own and the dominant firm's fitness while the latter behaves individualistically.

The origin of these different selection criteria (agent goals) can be justified by different forms of co-ordination in the cluster. In the existing literature, the mode of co-ordination in clusters is a consequence of the local institutional infrastructure, the so-called *local culture*. The nature of such local cultures can however differ both with respect to the mindset and agent behaviour they favour as well as in the context of their sustaining mechanism (power distribution).

In the ideal-typical case of clusters with even power distributions, agents adhering to an informal institutional infrastructure may exhibit very collective mindsets.

At the same time, local cultures in even-power clusters might also lead to very individualistic orientations. Dropping the assumption of even power distributions in the cluster then enables the emergence of alliance or leader firm scenarios. If cluster agents form more closely tied business groups (as was found in many Italian Industrial Districts) it is likely that the goals underlying their respective strategies will prioritise the alliance over other cluster agents. This is approximated in the model by assuming that alliance partners seek to maximise the joint fitness of themselves and their ally. In the case of a leader firm in the cluster, this agent is in a unique power position implying that it can use this influence to gear the strategy choices of other agents. The remaining cluster agents (subordinates) would therefore attempt to select strategies that benefit themselves and the leader firm in order not to mess with the biggest power in the cluster. The leader firm in turn is free to choose any strategy it sees fit, i.e. it can care only for its own fitness.

Note that (especially in even-power clusters) the model does not describe specific institutional setups leading to these different mindsets, but only their overall result: There are many possible ways of ensuring that actors care for themselves or the cluster as a whole. Collectivism for instance might result from a (culturally induced) lack of egoism, strong community and family ties in the cluster as well as a strong threat of being shunned or sued by other agents when behaving egoistically (Holländer 1990).

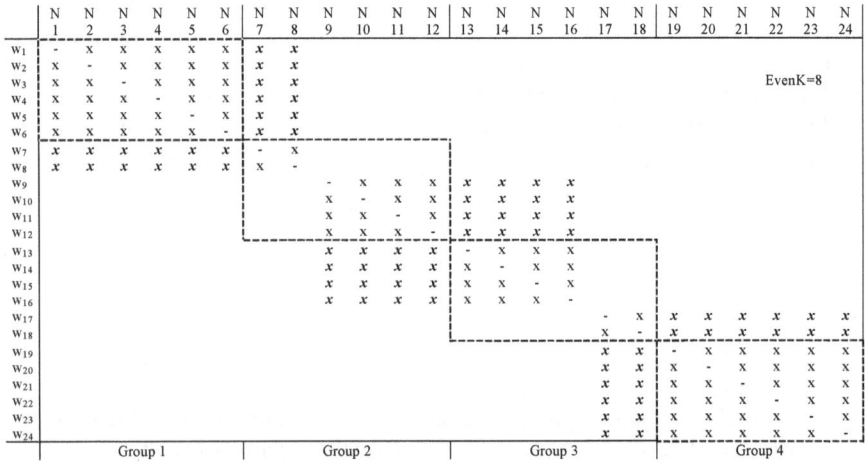

Fig. 7.6. Search (- -) and selection (___) landscape in individualistic clusters

The role of co-ordination as a variable influencing agent's strategy selection can be compared to a *filtering* function. While all agents are able to generate the same number of modifications through their identical search process (described in Sect. 7.1.3), the number of accepted modifications will differ according to the co-selection mechanism employed. In the case of individualistic clusters, agents search and evaluate strategies based on the fitness of the elements under their control. These strategies are sought and chosen on the same fitness landscape (Fig. 7.6.): That of the agent's substring containing n=6 elements. Depending on the extent of division of labour, this fitness landscape exhibits different levels of K in-terdependencies and is deformed to differing extents by the activities of other agents (depending on the C externalities). As agents care only about their subset of the fitness landscape, individualistic clusters do not internalise any C externalities into agent decision making.

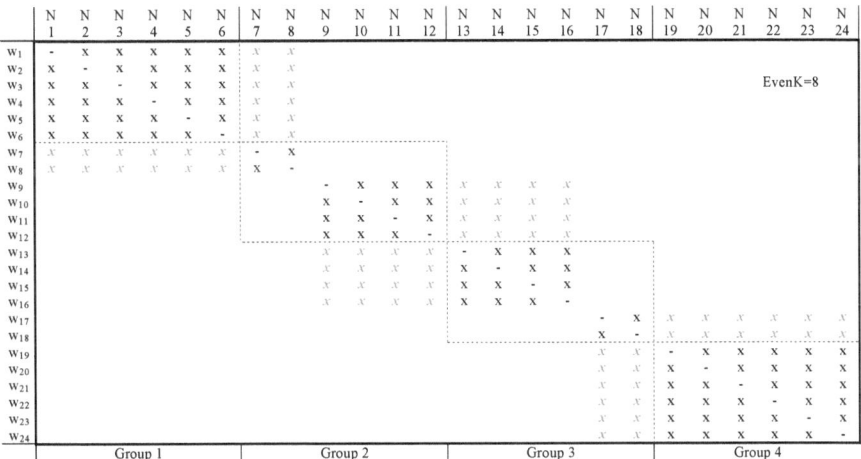

Fig. 7.7. Search (- -) and selection (__) landscape in collective clusters

At the other extreme, agents in collective clusters aim at improving the average fitness of all $N=24$ elements through their search activity within a subset of the fitness landscape. Elements *outside* an agent's control therefore enter the evaluation of her modifications, which constitutes an extension to the N/K literature. To date, most of the literature assumed an identity of the elements used in search and selection processes. This corresponds to the individualistic case where agents search over the six elements within their control and evaluate any modification accordingly. Introducing elements outside the jurisdiction of the agent into her selection criterion implies that search is conducted over $n=6$ elements but modifications are only accepted if the average fitness of all $N=24$ elements is improved.[113] As a result, a number of modifications based on exploring the $n=6$ substring will be rejected by cluster agents as they are unlikely to meet the selection criterion. The number of selected agent modifications will be lowest in the case of collective clusters as agents search on a $n=6$ element landscape but evaluate strategies on an $N=24$ element one. By integrating all system elements into each agent's selection mechanism, this mode of co-ordination internalises all C externalities (highlighted by the grey 'x' in Fig. 7.7.). Thus, modifications selected by agents will be better for the system: While unexpected (adverse) C effects due to simultaneous strategy changes of several agents can still occur, agents try to anticipate the effects of their activities on all others before selecting a strategy.

---

[113] In this setting, agents are willing to accept decreases in their own fitness as search is not pareto-optimal.

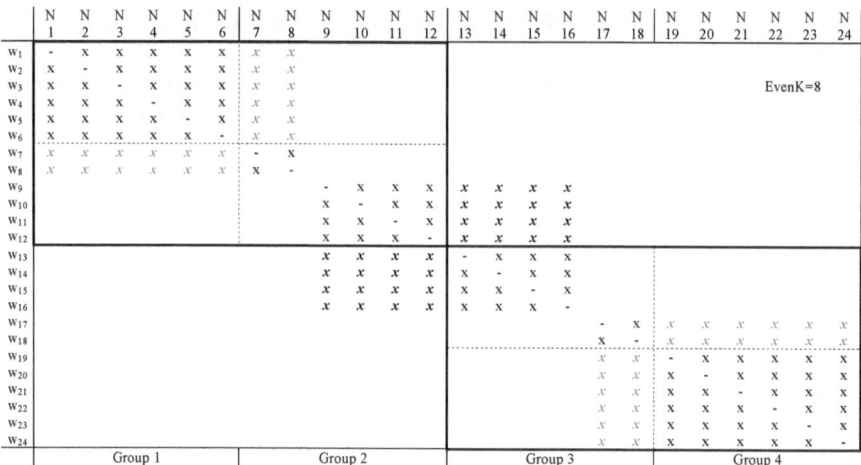

Fig. 7.8. Search (- -) and selection (__) landscape in alliance clusters

In the case of alliances, agents aim at improving the joint fitness of themselves and their partner. As a result, modifications are sought on a n=6 landscape and agents evaluate them on an n=12 one. Fig. 7.8. depicts the search and selection landscapes in an alliance cluster with partnerships between groups one and two as well as three and four. Depending on the extent of division of labour as well as the alliance constellation, this co-ordination mechanism internalises a given number of C externalities (in grey). Remaining interdependencies between both alliances (represented by bold 'x') imply that this co-ordination mechanism can yield sub-optimal results from a system's (cluster) perspective (see Chap. 6).

In clusters with a dominant firm, all agents gear their activities at improving their own and the dominant firm's fitness whereas the latter behaves individualistically (Fig. 7.9.). Three agent groups search modifications on a n=6 element fitness landscape and select them on a n=12 element one. The dominant firm in turn searches and selects modifications based on an n=6 element fitness landscape. This co-ordination mechanism internalises all C externalities that other cluster agents exert on the dominant firm (see 'x'). It does not internalise any C externalities caused by the dominant firm, or any interdependencies between subordinate agent groups (represented by 'x' in Fig. 7.9. This mechanism can thus lead to suboptimal results for the system. Moreover, the behaviour of the dominant firm is decisive: With respect to their search and selection mechanism, the other cluster agents behave similarly to the alliance scenario *if they generate externalities on the dominant firm*.[114] Consequently, the performance of the dominant firm cluster

---

[114] If no externalities exist between agents, their search behaviour will always generate improvements even if activities outside the agent's control are taken into account. Agents thus effectively behave like in the individualistic case. This leads to significant instability for leader firm co-ordination mechanisms if EvenK is low: Agents in groups that are more distant to that of the dominant firm are not connected to it and thereby behave individualistically. A change in this phenomenon can be expected once externalities EvenK increases.

as compared to the alliance case depends on the extent to which externalities between subordinate agents and the leader firm exist (i.e. on the centrality of the latter) as well as on the number of disturbances the dominant firm generates through its individualistic behaviour (a direct consequence of the K value in its fitness landscape subset).

	N1	N2	N3	N4	N5	N6	N7	N8	N9	N10	N11	N12	N13	N14	N15	N16	N17	N18	N19	N20	N21	N22	N23	N24
W1	-	x	x	x	x	x	x	x																
W2	x	-	x	x	x	x	x	x																
W3	x	x	-	x	x	x	x	x													EvenK=8			
W4	x	x	x	-	x	x	x	x																
W5	x	x	x	x	-	x	x	x																
W6	x	x	x	x	x	-	x	x																
W7	x	x	x	x	x	x	-	x																
W8	x	x	x	x	x	x	x	-																
W9									-	x	x	x	x	x	x	x								
W10									x	-	x	x	x	x	x	x								
W11									x	x	-	x	x	x	x	x								
W12									x	x	x	-	x	x	x	x								
W13									x	x	x	x	-	x	x	x								
W14									x	x	x	x	x	-	x	x								
W15									x	x	x	x	x	x	-	x								
W16									x	x	x	x	x	x	x	-								
W17																	-	x	x	x	x	x	x	x
W18																	x	-	x	x	x	x	x	x
W19																	x	x	-	x	x	x	x	x
W20																	x	x	x	-	x	x	x	x
W21																	x	x	x	x	-	x	x	x
W22																	x	x	x	x	x	-	x	x
W23																	x	x	x	x	x	x	-	x
W24																	x	x	x	x	x	x	x	-
	Group 1								Group 2				Group 3						Group 4					

Fig. 7.9. Search (- -) and selection (__) landscape in leader firm clusters[115]

Similarly to the notion developed in Sect. 6.1.3, any mode of co-ordination extending the evaluation landscape of agents (i.e. collective, alliance and leader firm) beyond the elements under agent control faces a trade-off as compared to the individualistic case. While the individualistic mode generates improvements at the agent level very quickly, the latter may be sub-optimal from the cluster's perspective (depending on the degree to which externalities between agents exist). At the same time, more collective modes of co-ordination take longer to generate modifications at the agent level (especially as the size of the evaluation landscape increases). The modifications found will however be better from the system's perspective. Trade-offs can therefore emerge with respect to speed and performance of cluster adaptation. The smaller likelihood of selection for co-ordination mechanisms with more collective mindsets implies slower adjustment processes. At the same time, it conveys a stability benefit for the co-evolving system cluster: Fewer changes in agent and group configuration lead to fewer deformations between landscape subsets. Moreover, the speed disadvantage of non-individualistic clusters is smaller than in the existing N/K literature. As agents explore a search space of n=6 elements while evaluating strategies on an n=12 or n=24 landscape, they will be faster in encountering superior configurations than agents exploring n=12 or n=24 elements with the same search process.

---

[115] Leader firm located in group 1.

By holding agent search and group selection (bidding) processes identical for all cluster types, an isolation of the roles of division of labour and mode of co-ordination becomes possible. The division of labour will feed back onto agent behaviour by determining the structure of their fitness landscape subsets, the extent to which agent activities deform the fitness landscape of other agents as well as the centrality of different cluster actors. The mode of co-ordination in turn influences the stability of adaptation processes as well as the optimality of results from a system perspective. The more C externalities are included in agent decision-making, the fewer modifications get to be selected by individual actors. This generates more stable adjustment processes with better results for the entire system.

7.3 Number, optimality and spread of modifications: Driving adaptability

Within the N/K framework, the adaptive performance of clusters can be measured in two ways. One involves the level of fitness obtained through agent-driven adaptation. Another regards the fluctuation in cluster (system) fitness: Since clusters compete in a wider environment, very drastic falls in fitness would affect their likelihood of survival. In fact, the gains from a higher fitness value obtained throughout the adaptation process might even be offset by the losses incurred if fitness fluctuates (falls) substantially during that time. As a result, both average fitness and standard deviations were gathered in the simulation analysis. Two separate sets of hypotheses on the effects of structure (division of labour) and selection (co-ordination) for cluster adaptation to change events can be derived from the explanations provided in the previous chapter (regarding general N/K dynamics) as well as Sect. 7.2. They can be based on the role of division of labour and co-ordination for the number, optimality and spread of selected agent modifications. While these three factors are reciprocally interdependent, adopting an "all else held equal" perspective leads to the following conclusions.

In interdependent systems where different agents and agent groups strive for better configurations, the *number of modifications* effected by each agent group will play a crucial role for the stability of adaptation at the system level. If each group in the cluster generates a lot of modifications, these will have repercussions on other agent groups which then in response change their configuration. This leads to an ongoing dance where agent groups keep altering their configuration in response to one another and thus decreases the stability of adaptive performance for the system as a whole. The number of selected modifications in each agent group then depends on the structure of the fitness landscape in the group's substring (K values) as well as the number of agents per group.

A very closely related aspect is the *spread of agent group modifications* within the system. As interdependence between system groups increases, even small changes in any of them have far reaching consequences within the system and would provoke responses from a greater number of other groups. With growing levels of inter-group linkages in the system, stability of adaptation therefore declines.

While adaptive performance measured by average fitness automatically declines as fluctuation in adaptive processes increases, a third aspect matters more specifically for the quality of the results obtained by agent and group dynamics, *namely the system-level optimality of agent modifications*. As was already outlined in Sect. 6.1.3, the selection of modifications faces a trade-off. The smaller the number of elements used to evaluate and select changes in agent configuration, the faster improvements are encountered. However, if interdependencies in the system go beyond the elements used for strategy evaluation, this implies that the selected modifications can be sub-optimal from a system perspective, which would lead to lower adaptive performance in terms of average system fitness.

All three aspects are influenced by the extent of division of labour and the mode of co-ordination in clusters, which allows for a formulation of hypotheses of how both factors may matter for performance and stability of cluster adaptation. The division of labour affects adaptation through two contradictory forces. On the one hand, a greater extent of division of labour (mirrored by higher EvenK values) increases the interdependence within the substring controlled by each agent (K). As a result, individual agent search activity is expected to lock-in to a local optimum more quickly. At the same time, increases in EvenK always imply growing cross-agent interdependence (C) mirroring agglomeration externalities. In consequence, even small changes in agent configuration have strong repercussions on the fitness landscapes of other agents, leading to greater fluctuation in adjustment processes.

Taking the effect of limited element interdependence on average fitness values in the landscape into account (see Sect. 6.1.1), an optimum degree of division of labour is expected to exist in clusters. With some interdependence between elements (small C and K values), a growing number of random draws for each element's conditional fitness contributions increase average fitness of landscape optima. A complexity of the cluster's fitness landscape growing beyond this point however leads to a lower performance in adaptation because "*in functioning complex systems with many highly differentiated and tightly interdependent parts, it is highly unlikely that undirected change in a single part will have beneficial effects on the system*" (Nelson and Winter 1982, p. 116). As a result, very interdependent clusters (high values of K and C) will fare worse in adapting to change both in terms of average fitness as well as fitness fluctuation in adaptation.

Proposition 1. Beyond a threshold value, higher system interdependence (K/ C), leads to lower adaptive performance regarding average system fitness and standard deviation.

The effect of selection (or co-ordination) for system adaptability will depend on three aspects: The existence of inter-agent externalities, the extent to which selection accounts for them as well as the overall structure of the fitness landscape. The first point refers to the fact that selection criteria enabling some sort of co-ordination between agents will only become relevant if the latter's fitness is interdependent, i.e. if there are C externalities between actors. If such cross-agent interdependencies exist, those agents co-ordinating their activities fare better than those acting egoistically, especially as the role of C externalities increases.

Proposition 2. The performance of different co-ordination mechanisms will differ more strongly as the role of cross-agent externalities (C) increases.

In the presence of cross-agent externalities, the quality (performance) of any co-ordination mechanism then crucially depends on the extent to which it captures the latter, i.e. whether or not the 'right' agents are co-ordinating their activities. Put differently, system performance will increase, the greater the share of C externalities captured by the co-ordination mechanism.

Proposition 3. The greater the internalisation of C externalities through the co-ordination mechanism, the higher the system fitness.

On a final note, the stability of cluster adaptation will also depend on the number of agents in each group controlling a substring of the system. Since agents are chosen to represent their group based on the cluster fitness offered by their configuration (see Sect. 7.1.4), a greater number of agents per group implies that competition for selection at the substring level is greater: If there are less agents in each group trying new configurations of their elements, experimentation shifts from the group to the system level. Configurations that would not win the bidding process in larger agent populations could come to be the best configuration obtainable. As a result, the fluctuations of system fitness increase. The following chapter investigates these propositions by presenting the simulation results for clusters adapting to bifurcation and perturbation events. While absolute values of both are not comparable (as the systems walk different fitness landscapes), relative performances yield a number of insights.

Part V Division of labour, co-ordination and cluster adaptation

Every concrete process of development [...] rests upon preceding development. [However, sometimes a system experiences a change] which so displaces its equilibrium point that the new one cannot be reached from the old one by infinitesimal steps. Add successively as many mail coaches as you please, you will never get a railway thereby.

Joseph A. Schumpeter 1934a, p. 64

Conducting simulations and comparing adaptive performance of clusters differing in given architectures yields that division of labour and mode of co-ordination matter for adaptation by influencing the number, system-level optimality and spread of agents' modifications. There is an optimum degree of division of labour maximising cluster adaptability while more collective modes of co-ordination (collective, alliance) usually outperform more individualistic ones (leader firm, individualistic). Moreover, clusters with greater numbers of agents per group perform better in adaptation thanks to a greater opportunity for experimentation and diffusion of best practice among group agents.

The derived causalities contribute to an explanation of empirical realities. The move of Italian Industrial Districts towards alliance and leader firm structures results from two different incentives in each agent group. First, any group can benefit from becoming dominant within the cluster as this maximises group fitness. Second, group fitness exhibits a prisoner's dilemma payoff structure in the sense that each group benefits most when it can behave individualistically in an otherwise collectively oriented cluster. It suffers worst if the neighbour group behaves individualistically. Alliance and leader firm can thus be good intermediate solutions in the face of instable collective agent orientations.

Regarding the Silicon Valley – Boston 128 comparison, the greater number of agent numbers in the Valley as well as a possibly more collective co-ordination mechanism speak in favour of its adaptive performance while the Boston 128 exhibits advantages regarding a lower degree of cross-agent externalities. This phenomenon however requires further elaboration. Taken together, the study shows that cluster development is a path dependent process: Architectures having evolved throughout the cluster's history may or may not help its agents accommodate for and thereby survive upcoming change events. The findings of this study could therefore have further implications for future empirical work, policy initiatives and especially theoretic research.

8 Clusters, change and adaptation – Simulation results

The simulation model developed in Chap. 7 is a tool for analysing the roles of division of labour and co-ordination in cluster adaptation to change events. This aim is achieved by building clusters with different given architectures that might have evolved throughout the development preceding the change event in each case. The model then simulates and compares the adaptive performance of these clusters both with respect to one another as well as between different degrees of fitness landscape complexity (different parameter settings). Adaptability then relates to the system's ability to discover good (fit) configurations following a change event.

Comparing adaptive performance of clusters over different parameter settings accounts for the role of the extent of division of labour. The simulations find (in line with previous research on N/K systems; see Sect. 6.1.1) an optimum degree of interdependence (complexity) for all modes of co-ordination. In the context of clusters, this means that there is an optimum degree of division of labour that maximises adaptive performance. Comparing adaptation in clusters with different architectures regarding agents' selection mechanisms gives an account of the role of co-ordination mechanisms. It is found that collective clusters where agents choose their strategies for the good of the cluster as a whole perform best in adaptation whereas individualistic clusters composed of egoistic agents do worst. Alliance and leader firm constellations position themselves between these two extremes, shifting in relative performance as cross agent interdependencies (C) increase.

Both types of results are robust with respect to test simulations altering the number of agents in the cluster: While relative performances with respect to modes of co-ordination and degree of division of labour remain the same, absolute cluster adaptability increases with the number of agents. This effect is caused by the fact that a greater number of agents allows for more experimentation at the group level. If the best alternatives from this experimentation process are selected to represent groups at the cluster level, more experimentation leads to better results from the cluster's perspective. Furthermore, findings hold regardless of whether the environmental change challenging the current cluster configuration comes in the guise of a bifurcation or a perturbation event.

The strong performance of collective clusters brings up the question why many real world cases move towards alliance or leader firm scenarios. To address this issue, the performance of all modes of co-ordination is also investigated for the group level. It could be imagined that clusters shift in co-ordination mechanism because the mode that is best for the system as a whole does not perform too well for agent groups. Except for the case in which an agent group is the leading firm, however, the collective case is the best mode with respect to adaptive performance both at the system as well as the agent group level (with very few exceptions). This aspect alone can therefore not explain shifts in the mode of co-ordination in clusters.

The collective cluster can however be destabilised by the emergence of egoistic agent groups. Comparing the fitness of egoistic and collectively oriented agent groups in clusters shows a strong prisoner's dilemma: While any agent group benefits most from being the sole egoist in the cluster it also suffers worst if its neighbour (the group it is most linked to by agglomeration externalities) decides to behave egoistically. The fitness of the agent group in this case is lower than if all agent groups behave individualistically. As a result, having one egoistic agent group in the cluster is sufficient to destabilise the collective mode of co-ordination. Unless agent groups are independent, egoistic behaviour spreads between neighbours and thus throughout the entire cluster. This phenomenon only occurs if all group agents behave egoistically implying a greater vulnerability for clusters with smaller numbers of agents per group. Again, the results are identical for bifurcation and perturbation events. The implications of these findings will be discussed in more detail in Chap. 9.

Simulations were run with nine populations representing differently configured clusters (with respect to their co-ordination mechanism) for increasing values of EvenK representing a growing division of labour. Since it emerged that the relative performance in those populations representing different alliance (3-5) or leader firm constellations (6-9) was exclusively related to their ability to internalise C externalities, only the results for the best performing cases will be given here. That is, alongside the cases of individualistic (1) and collective (2) clusters, results are presented for the best performing alliance (1+2, 3+4 (4)) and leader firm scenario (leader firm in first group (6)). For other results see appendix 6.

8.1 Fitness landscape reconfigurations (bifurcation)

Simulations representing cluster adaptation to bifurcation events were run for 600 steps every time. In order to arrive at general results, 100 simulations were conducted for each EvenK parameter value. The complete alteration of the fitness landscape that was argued to occur in the face of bifurcation events was introduced by starting a new simulation with newly configured agents on a new fitness landscape. This approach was chosen for two reasons. First, its effect is very similar to keeping the current agent configurations and altering all element fitness contributions: The fitness of agents and clusters changes very strongly in both cases. Second, representing bifurcations like very large-scale perturbations (i.e. conducting one simulation of 60,000 steps with a shift in the fitness landscape affecting all N=24 elements every 600 simulation steps) was technically unfeasible. This approach would already lead to system crashes (due to memory limitations) if the shift in the landscape affected only six elements (see also Sect. 8.2). It would only have increased with the number of elements altered by the change events, thereby narrowing down the parameter scope of analysis to an unacceptable extent.

In order to allow for a comparison of adaptation outcomes in clusters, performance (average fitness) and stability of adaptation (standard deviation of fitness values in each simulation) were obtained at the level of agent groups and for the entire cluster following each simulation run. These results were then averaged over the 100 rums conducted with each parameter value. Simulation runs included parameters EvenK=4 up to EvenK= 24 with the exception of EvenK=2; 3 and 6. The neglect of these parameter values is owed to the fact that the fitness landscapes characterised by interdependent element blocks of size two, three or six do not exhibit any cross-agent externalities (C) given the landscape decomposition of n=6 elements (out of 24) per cluster agent.[116] However, C externalities are crucial to the existence of clusters as they are defined here. If there are no benefits to having a number of independent firms in an area, i.e. if no agglomeration externalities exist, an integrated firm can be as efficient as a cluster of the same size. Clusters therefore can only be represented in fitness landscapes exhibiting interdependence at the agent level (K) as well as between agents (C).

What emerges very clearly is a trend of increasing chaos in system behaviour as landscape complexity (K, C) increases, i.e. fluctuations in system fitness throughout adaptation processes increases markedly. Fig. 8.1. depicts two example runs of four clusters with different co-ordination mechanisms (individualistic, collective, alliance and leader firm) that adapt to bifurcation events at simulation steps 600 and 1200. For low landscape complexity (EvenK=4), each bifurcation is followed by relatively swift adaptation resulting in stability of system configurations. As landscape complexity increases, however (EvenK=14) system dynamics following a bifurcation are far less stable (especially for individualistic and leader firm clusters). These example runs already indicate an overall trend in the simulations where growing landscape complexity leads to greater standard deviation and lower mean fitness values, i.e. decreasing performance and stability of cluster adaptation.[117]

---

[116] The parameter value of EvenK=1 represents an independent system where each element influences only its own fitness contribution. This case does not apply to the analysis conducted here either.

[117] In fact, system dynamics are chaotic for most co-ordination mechanisms (except the collective case that *always* finds stable solutions) once EvenK increases beyond 12.

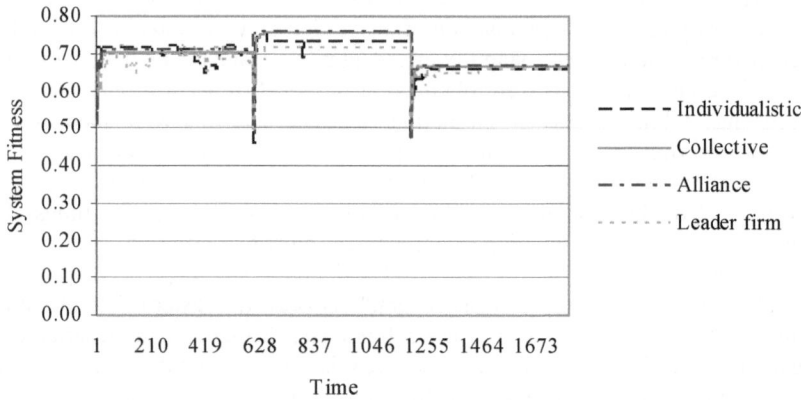

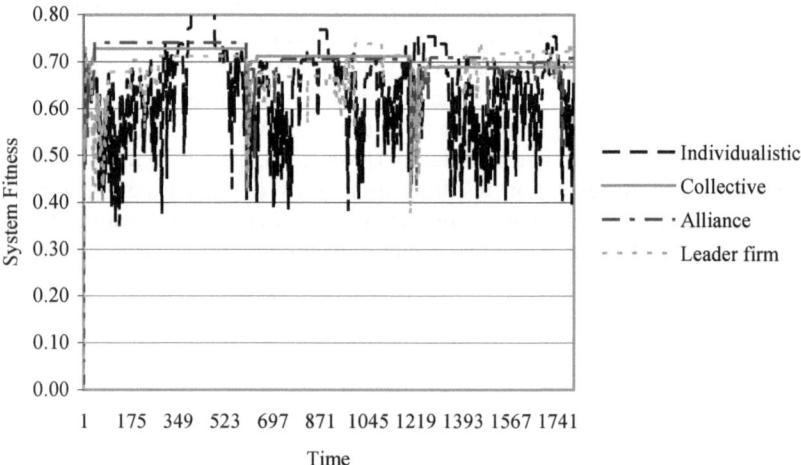

Fig. 8.1. Landscape complexity, bifurcation (steps 600 and 1200) and cluster adaptation - Example runs for EvenK=4 and EvenK=14

With respect to the role played by the extent of division of labour for cluster adaptation, Fig. 8.2. highlights a trade-off between division of labour and adaptability. Regardless of the mode of co-ordination in clusters, the simulations point at an optimum degree of division of labour where the average fitness obtained in adaptation processes is maximised. This occurs when C=3.0 (corresponding to a parameter value of EvenK=9). Beyond this point, adaptive performance decreases for all forms of co-ordination. This observation is in line with proposition 1 stating that *average system fitness decreases and fluctuation in adaptation process increases with growing landscape complexity.*

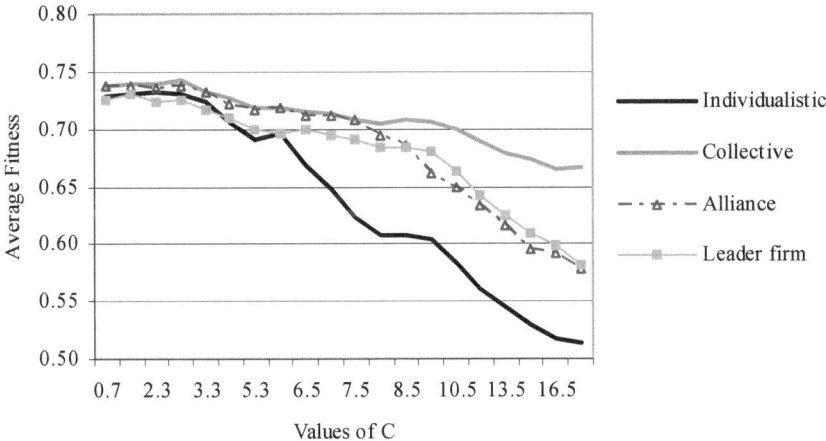

Fig. 8.2. Division of labour, co-ordination and adaptive performance

This phenomenon is attributable to the properties of N/K fitness landscapes: For small levels of interdependence, fitness contributions of individual elements start being drawn randomly more often than for almost independent cases. As a result, there is a certain probability that average fitness contributions for each element (taking all conditional fitness values into account) will be higher than in the independent case where the element fitness value is only drawn once. Moreover, in landscapes with intermediate complexity, the issue of conflicting constraints in element optimisation does not materialise too strongly as interrelations between the former still tend to be limited (see also Sect. 6.1.1). As landscape complexity increases, however, intra and cross-agent interdependence grows. The increase of cross-agent externalities then implies that small changes in one agent group configuration spread far in the cluster, thereby inducing a response by other agent groups. In consequence, in very interdependent clusters, agents disturb each other's fitness landscape more strongly. Furthermore, with growing element interdependence, the issue of conflicting constraints between elements can outgrow the benefits of multiple draws of element fitness contribution. For both reasons, very interdependent clusters fare worse in finding stable and good solutions in their adaptation process. The exact value of the optimum degree of division of labour is then owed to the properties of fitness landscapes in the N/K model.

A less obvious aspect uncovered by the simulations is the impact of different co-ordination mechanisms on cluster adaptation. This role of co-ordination for adaptive performance can be linked to three different aspects: The extent of interagent externalities (C), their internalisation through different co-ordination mechanisms as well as the latter's effect on stability. The first aspect is related to the observation that the loss in adaptive performance with increasing landscape complexity differs between the modes of co-ordination. For small values of C, all decision-making mechanisms perform relatively similar. Beyond the optimum value C=3.00, a gap begins to open up. The *collective* cluster performs best by losing only 9.59% of its fitness between the simplest and the most complex fitness landscape. The *individualistic* scenario performs worst and loses almost one third (29.55%) of its fitness. Both *leader firm* and *alliance* scenarios position themselves between these two extremes, losing about 20% of fitness as landscape complexity becomes maximal. This observation is in line with proposition 2: *Co-ordination begins to matter as the role of cross-agent externalities (C) increases.*

Second, throughout all simulations, the collective cluster performs best both in terms of maximising average fitness as well as minimising system-wide fitness fluctuations (Fig. 8.3.). The individualistic cluster in turn begins faring worst once the importance of inter-agent externalities increases (beyond EvenK=10/ C=4.04). An interesting phenomenon lies with the relative performance of alliance versus leader firm clusters: As C increases, the authority scenario involving a central leader firm begin to perform better than alliance ones, both in terms of average fitness as well as its standard deviation. This shift in performance occurs once C>9 (EvenK>18) for the best performing alliance and leader firm scenario.[118]

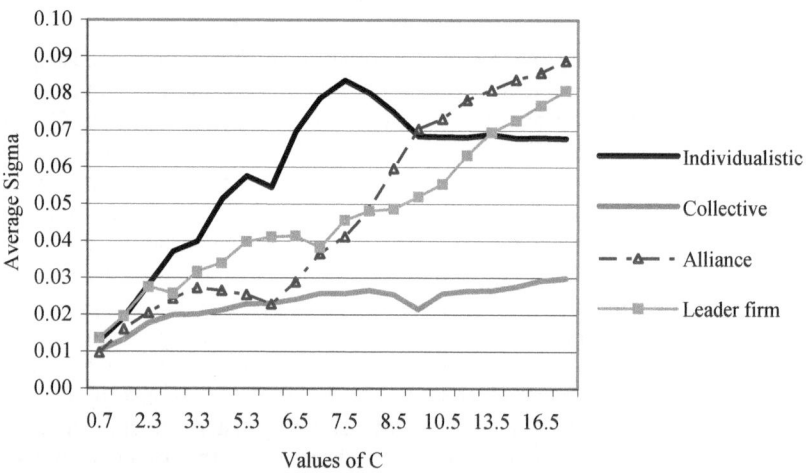

Fig. 8.3. Division of labour, co-ordination and stability in adaptation

The differences in performance and stability of cluster adaptation processes (according to their degree of division of labour and co-ordination mechanism) can be explained by a trinity of aspects regarding the number, optimality and spread of

---

[118] The best performing leader firm cluster starts outperforming inferior alliance scenarios far earlier than that (see also Fig. A6.1. in appendix 6).

individual agent modifications. As is found when investigating the role of division of labour for cluster performance, the more agents change their configurations and the greater the spread of these changes within the system, the worse adaptive performance becomes. The number of selected agent modifications then depends on structure and also selection aspects. The structure of the system landscape subset (extent of K) determines the likelihood of lock-in to a (local) optimum for any agent. All else held equal, increasing values of K therefore stabilise agent activity. However, due to the block distribution of interdependencies in the LSD programme, increases in fitness landscape subset interdependence (K) occur alongside increases in cross-agent externalities C, i.e. even small changes in agent activity have large repercussions on other agents implying that adaptive performance decreases and fluctuations grow.

For any *given* degree of landscape interdependence, the selection process of agent strategies that reflects the different modes of co-ordination begins to play crucial role. As was elaborated in Sect. 7.2.2, selection acts as a 'filter' in agent behaviour. It determines the *number* of modifications that will be executed at the agent level on the one hand as well as the *quality* of these modifications from the system perspective on the other. With respect to the number of selected modifications, the following aspects hold. All agents are able to generate the same number and kind of changes to their configurations (due to their identical search process). Depending on their selection criteria, a number of these modifications are however rejected. With more collective orientations, the likelihood of finding better configurations and thus the number of *accepted and executed* (selected) agent modifications decreases. As an example, agents in individualistic clusters search and evaluate their strategies on an n=6 landscape. In this context, the likelihood of finding improvements in the configuration and thereby the number of selected modifications is greatest. In the collective case, all agent search modifications on an n=6 landscape while evaluating them on an N=24 one (evaluation is based on overall system fitness). The likelihood of finding improvements to the system landscape based on an n=6 subset is low compared to the individualistic case, i.e. the number of selected modifications is lowest in collective clusters.

In alliance clusters, all agents search on an n=6 landscape and evaluate their choices on an n=12 one. Again, the likelihood of finding improvements to this bigger landscape is higher than in the collective and lower than in the individualistic case. The number of selected modifications in alliance clusters corresponds to this. Similarly, in the leader firm case, all subordinate agents search on n=6 and evaluate modifications on an n=12 landscape, i.e. the number of their accepted and executed modifications corresponds to the alliance case. The difference between the two constellations then lies with the behaviour of the leader firm which searches and evaluates strategies egoistically on an n=6 landscape and thereby generates a greater number of modifications. All else held equal, the following ranking of co-ordination mechanisms with respect to the number of modifications at the agent level can be derived: *Individualistic> Leader firm> Alliance> Collective*. Depending on the extent of C externalities, this greater number of modifications then spreads within the cluster meaning that agents disturb each other's search process more or less frequently according to the co-ordination mechanism employed. This in turn impacts on the number of modifications executed in response. All else held equal, clusters with a greater number of agent-level modifications would therefore exhibit less stable adaptation processes.

A second role for the co-ordination mechanism in clusters arises with its impact on the *quality* of modifications from a system perspective. An optimisation of system fitness is only possible if all externalities between elements are taken into account (see also Sect. 6.1.3), i.e. the quality of results from a system perspective depends on the extent to which the different co-ordination mechanisms internalise cross-group externalities into the decisions of individual agents. As has been elaborated before (Sect. 7.2.2) the internalisation of C externalities is greatest in collective clusters and lowest in individualistic ones. Alliance clusters in turn rank before leader firm ones with respect to this factor as they mutually account for the C externalities between agent groups. In leader firm clusters, only those externalities conveyed on the leading firm are internalised. The quality of agents' modifications with respect to system fitness therefore ranks as follows: *Collective> Alliance> Leader firm> Individualistic*. Agents in clusters with more collective orientations execute fewer and better modifications in their configurations. In line with proposition 3, the *systems accounting for more C interdependencies exhibit better and more stabile adaptation processes*.

A puzzle in this context is posed by the shift in relative performance of alliance and leader firm clusters. Following the rationale developed above, agents in alliance clusters should find fewer and better modifications increasing average cluster fitness and decreasing the fluctuation in adaptive performance. However, the opposite occurs in clusters with very central leading firms (leader firm in the 1st to 3rd substring) once C>9 (EvenK>18). This is attributable to two special constellations in the fitness landscapes for these parameter values. First, as is apparent from Fig. 8.4., increases in the EvenK parameter imply an increasing centrality of groups one, two and to some extent three: These groups begin to receive and generate a greater number of C externalities. If the leading firm is located in the first group as presented here, increases in the EvenK parameter imply that other groups in the cluster are increasingly linked to the dominant actor by C externalities. For EvenK≤12, interdependence exists only between the leader firm (in group one) and the second group. For 12≤EvenK≤18, groups two and three are linked to the leader firm and for EvenK>18, all agent groups generate externalities on the leader firm and receive externalities from it.

This distribution of C externalities between agent groups matters crucially for the effectiveness of co-ordination mechanisms. It was argued before (see Sect. 7.2.2), that subordinate agents in leader firm clusters search and evaluate their modifications in principle identically to those in alliance clusters. Search takes place on an n=6 landscape whereas evaluation is based on n=12 elements. If there are no interdependencies between the elements affected by search and those on which evaluation is based, any improvement in the former will however be accepted. Put differently, agents in group 4 (in Fig. 8.4.) effectively behave like those in the individualistic cluster. Due to a lack of interdependence between search (n_{19}-n_{24}) and evaluation (n_1-n_6) elements, any improvement in the former will be executed. For small levels of fitness landscape interdependence, most agents in the leader firm case therefore effectively behave like in the individualistic constellation.[119] This leads to stronger disturbances between agent fitness landscapes as the *number* of selected modifications is higher than in the alliance case and their *quality* for the system is lower due to the more limited degree of C internalisation. For higher parameter values, the behaviour of subordinate agents increasingly approaches that of agents in the alliance case implying that the number of modifications decreases and their quality grows.

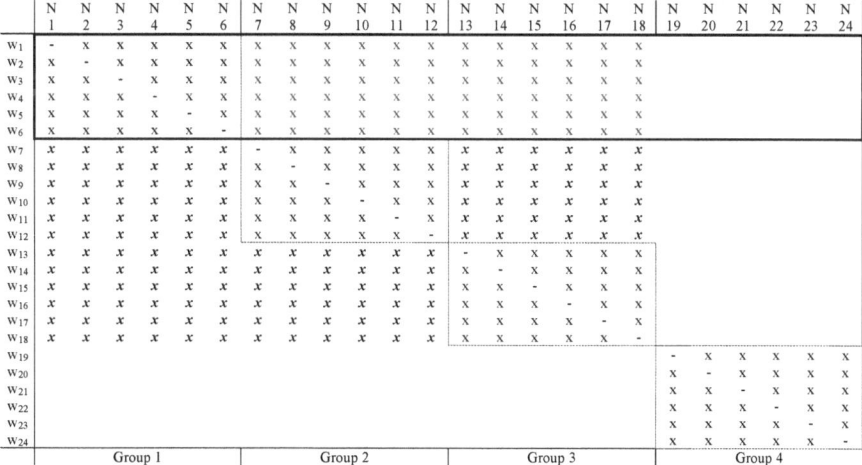

Fig. 8.4. Search and selection landscapes for leader firm clusters (EvenK=18)

Another aspect regards the behaviour of the dominant firm. As was established before (Sect. 7.2.2), it constitutes the key difference between the dynamics in alliance and leader firm clusters. Search and selection landscapes were very similar in both cases (n=6 and n=12 respectively) for allies as well as subordinate agents in the leading firm case. The individualistic orientation of the leader firm means that no externalities caused by that agent are internalised in its decision-making, thereby lowering the quality of leader firm modifications for the system as a whole. Moreover, it implies that the leader firm generates a greater number of modifications than other agents in the cluster (search and selection take place on n=6 landscape). This destabilises adaptation in leader firm clusters as compared to the alliance case since the leader disturbs the adaptation process of its subordinates to a greater extent than alliance partners do. While both concerns hold initially, with growing parameter values the fitness landscape subset for the leader firm grows more and more interdependent (see Fig. 8.4.). In consequence, it becomes locked-in to a local optimum more quickly, i.e. it is less active in changing its configuration and thus causes fewer disturbances in cluster adaptation.

---

[119] For EvenK≤12, groups 3 and 4 behave individualistically. For 12≤EvenK≤18, group 4 does and only for EvenK>18 do all subordinate agent groups effectively behave identical to the alliance cluster case.

Both effects work to offset the initial disadvantages of leader firm clusters as compared to alliance ones. What then causes their shift in relative performance is an additional effect that can be circumscribed as an *alignment of goals* in leader firm clusters. All subordinate agent search on an n=6 landscape and evaluate their modifications based on the aggregate of their own and the leader's fitness. As a consequence, they evaluate their changes on their n=6 subset and another subset consisting of the elements (n_1-n_6) controlled by the leading firm. This second subset is identical for all subordinates, i.e. their evaluations are subject to a partial alignment of goals. In comparison, alliance clusters base their strategy evaluation on separate subsets of the fitness landscape (n_1-n_{12} for groups one and two, n_{13}-n_{24} for groups three and four). With growing parameter values, the interdependence of elements between groups one, two and three increase. As a result, their effort to increase alliance fitness lead to a growing disturbance between them, i.e. there is no alignment of goals between alliances. This implies that the probability of finding stable configurations of the C related elements is lower for alliance clusters where agents try to optimise different yet connected subsets of the fitness landscape than for leader firm clusters where subordinate agents try to optimise the same subset of the fitness landscape and ignore the externalities between them. Thanks to the argument elaborated on above, this effect only materialises in leader firm clusters where the latter is a central actor that generates and receives a lot of C externalities. Having a leading firm that is otherwise unconnected to the cluster in terms of externalities leads to significant losses in adaptability as agents are gearing their activities in the 'wrong' direction (see results for population no. 9 in table A6.2. of appendix 6).

Alongside the cluster's co-ordination mechanism, the number and system-level optimality of agent modifications also depend on another aspect, namely the number of agents per group. Since individual modifications enter a bidding process at the group level, their number and optimality at the group level hinges on the number of modifications tried and evaluated there. If there are less agents in each group (A2(1) versus A10(1)), experimentation will shift from the group to the system level. Configurations that would not win the bidding process in larger agent groups could come to be the best configuration obtainable with fewer agents per group. As a result, the fluctuation of system fitness will increase (see also Figs. 8.5.-8.6.).

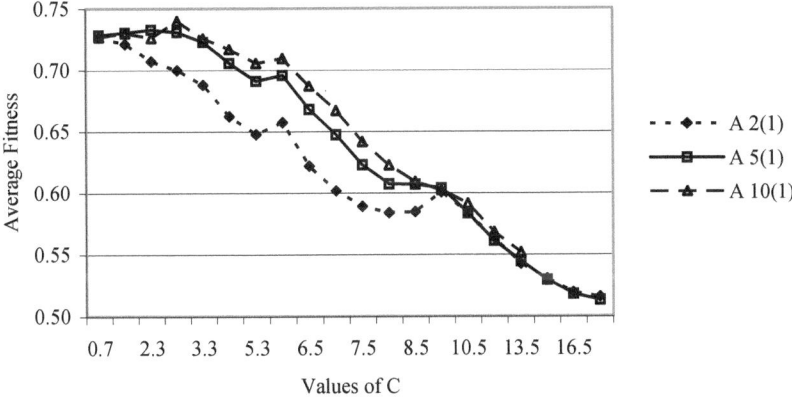

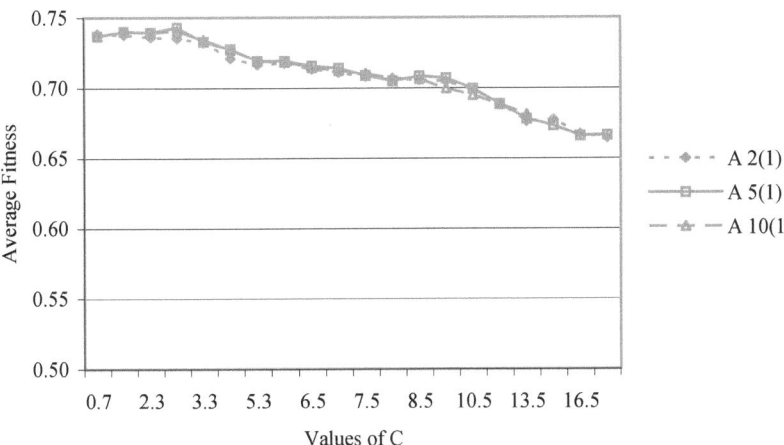

Fig. 8.5. Agent numbers and adaptation: Individualistic and collective [120]

Another issue regards the role of the number of agents in the leader firm scenario. Does its stability disappear, if a leading *group* of five agents replaces the leader firm (A5(5) versus A5(1))? It was found that this is not the case: The stability of the leader firm scenario in very interdependent clusters was attributable to the effects of C and K interdependencies outlined before. Both remain unchanged in the presence of a leader group. Instead, in line with the aforementioned reasoning on

---

[120] Results for the constellation with ten agents per group (A 10(1)) are only available for parameters up to EvenK=21, i.e. C=13.5. Simulations beyond that value lead to system crashes due to memory limitations.

the role of agent numbers, the existence of several leading agents decreases the fluctuations in this group's configuration, thereby exerting a stabilising effect on cluster adaptation.

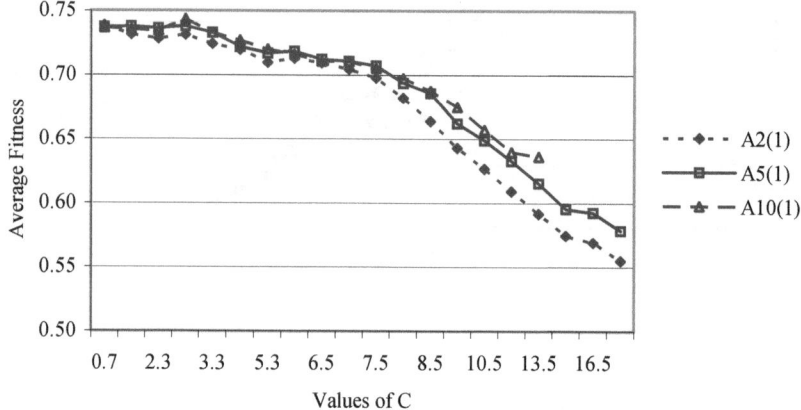

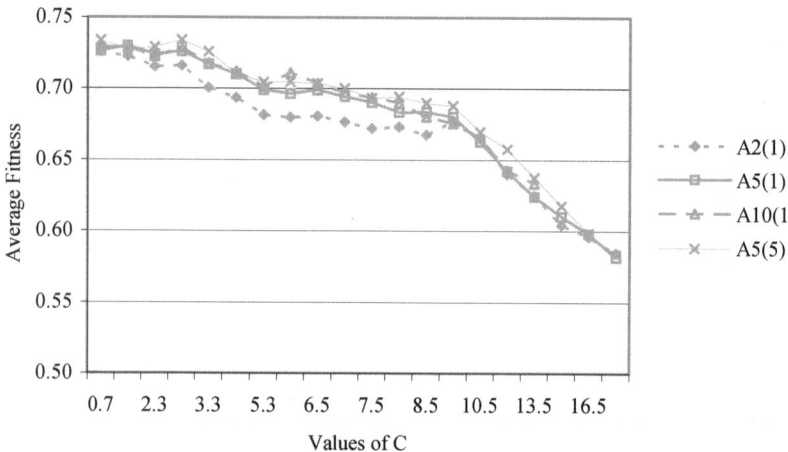

Fig. 8.6. Agent numbers and adaptation: Alliance and leader firm clusters

The extent to which the number of agents helps intra-group optimisation is also related to the frequency and optimality of modifications at the agent level. If individual agents generate and execute a wealth of modifications, having them compete in a larger group will significantly help system performance by weeding out more of those configurations that are bad for the system as a whole. This is attributable to the fact that the bidding process in agent groups is based on maximising expected cluster fitness (see Sect. 7.1.4). As a result, the impact of increasing numbers of agents is greatest for those co-ordination mechanisms in which individual agents encounter a greater number of modifications. Agent numbers thus matter most for adaptive performance in individualistic and leader firm clusters and only to a lesser extent in alliance and especially collective ones.

As landscape complexity increases, the impact of growing agent numbers on system adaptability becomes less pronounced in all cluster types. This is attributable to the fact that a greater interdependence of agent fitness landscape subsets (K) already constrains the number of agent modifications. This structural phenomenon materialises regardless of the co-ordination mechanism employed, i.e. as the C values increase, clusters with different sizes of agent groups once again start performing very similar in adaptation. In leader firm clusters, a leading group (A5(5)) increases performance more markedly than having ten agents in each subordinate group. This relates to the intuition developed before: The leader firm executes a greater number of modifications that are inferior for the system. It thereby has an adverse impact on cluster adaptation. If some of these modifications are eliminated due to a bidding process within a leading cluster *group*, these kinds of clusters become more adaptive to change events.

On average, clusters with more collective co-ordination mechanisms perform better in adaptation as their agents select fewer modifications that are better for the system. This effect materialises more strongly, the greater the spread of modifications in the cluster, i.e. the more cross-agent externalities exist. While all clusters lose in adaptive performance once the extent of division of labour passes a threshold value, this loss is lower under more collective co-ordination mechanisms (collective or alliance scenarios). Leader firm clusters can outperform their most closely related counterpart in agent dynamics (alliance clusters) only under very specific conditions in the system's fitness landscape: The cluster has to exhibit a very central and relatively inactive leader firm generating and receiving C externalities from and to all other cluster agents.

The relative differences in performance furthermore hold regardless of the number of cluster agents in each group. Absolutely, however, the number of agents per group increases adaptive performance, albeit to different extents for each co-ordination mechanism.

8.2 Fitness landscape shift (perturbation)

Simulations of adaptive performance in clusters facing a perturbation event were conducted over the same parameter range as in the case of bifurcations. Unfortunately, parameter settings of EvenK>17 lead to system crashes as the memory requirements became excessive. In consequence, results for parameters EvenK=18 to 24 are missing. This is however not too problematic since system behaviour in this parameter range tends to become very chaotic (see Fig. 8.7.), limiting the insights provided by results.

Within each simulation, cluster agents had to adapt to a perturbation event changing the fitness contribution of six randomly drawn elements (out of N=24).

Agents could then change and evaluate their configuration for 600 simulation steps before the next perturbation occurred. In total, simulations were conducted for 60,000 steps with a total of 100 perturbation events every 600 steps. This corresponds to agents in clusters adapting to a partially deforming fitness landscape for 100 times.

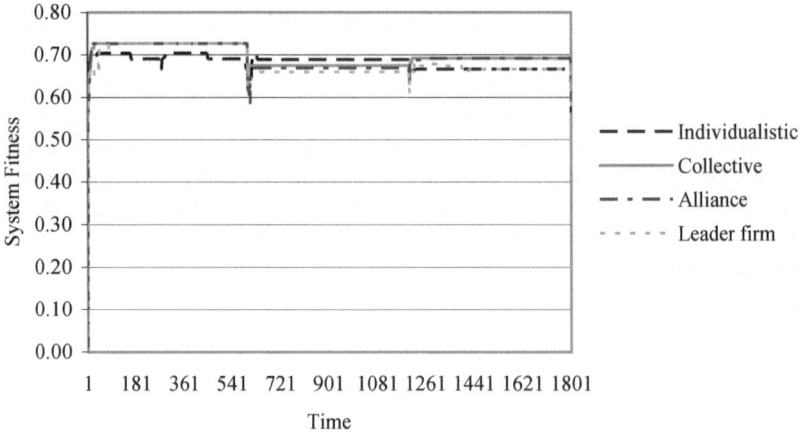

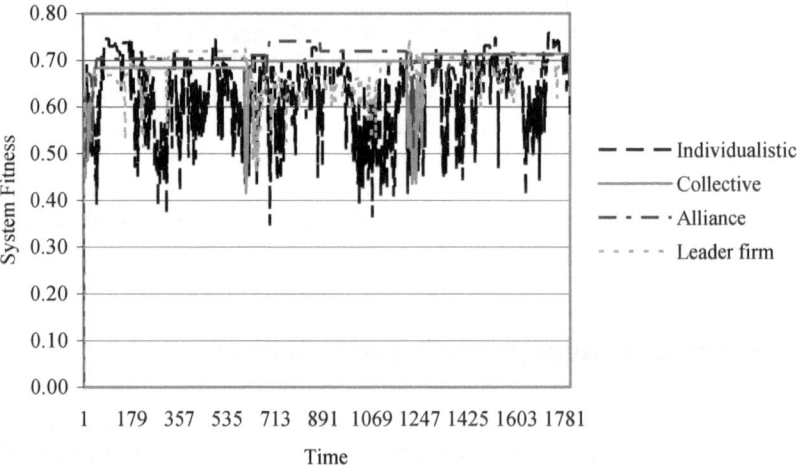

Fig. 8.7. Landscape complexity, perturbation (steps 600 and 1200) and cluster adaptation - Example runs for EvenK=4 and 14

Although the fitness landscape will change substantially between the first and the last perturbation, the fact that part of the landscape remains the same implies that one-off luck effects in adaptation may persist for a longer time in the presence of perturbations. Contrary to the idea of bifurcations where the entire fitness landscape is altered after 600 simulation steps, here only part of it is changed. If specific agents in a cluster were lucky in obtaining a good configuration prior to the perturbation event and if their subset of the fitness landscape remained unaltered by it, overall cluster performance in adaptation is influenced by persistence in fitness values of these lucky agent groups. However, the strength of this effect on adaptation performance is limited since perturbation events can affect different parts of the fitness landscape. Running simulations with a total of 100 such events restricts the effect of luck in agent adaptation. This factor also explains much of the similarity in results between the case of bifurcation and perturbation events, i.e. the drivers of adaptive performance work identically, regardless of whether all or parts of the fitness landscape are changed. Despite this, adaptation to perturbations is still easier for clusters as parts of the fitness landscape may be unaffected by the event. This allows for adaptation within a part of the cluster while maintaining the good configurations of agents in other subsets. On a case by case basis, fitness and thereby adaptive performance will thus be higher after perturbations.

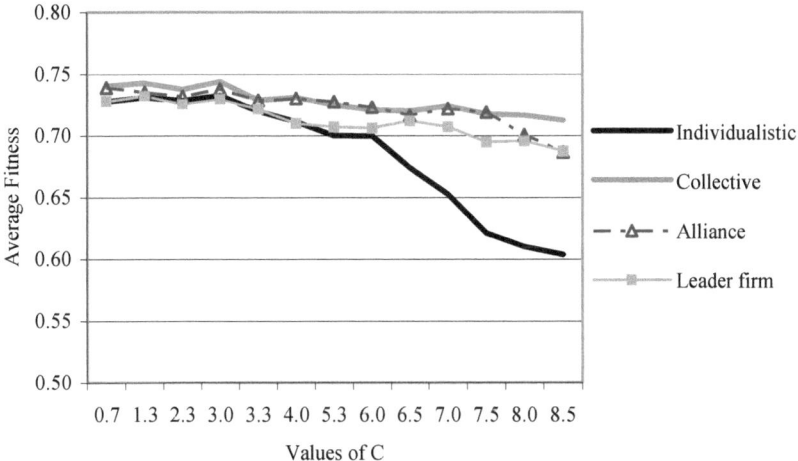

Fig. 8.8. Division of labour, co-ordination and adaptive performance

Comparing performance (average fitness) and stability of adaptation (average standard deviations of fitness values) for the case of perturbations yields similar results as in the case of bifurcations. Adaptive performance declines with the growing extent of C externalities once the latter has reached an optimum value (C=3.0/ EvenK=9; see Fig. 8.8.). This optimum degree of interdependence is the same as in the case of bifurcation events although one might intuitively have expected that smaller changes in the fitness landscape allow for a greater degree of division of labour while maintaining good adaptive performance. This intuition echoes the idea that clusters with a substantial degree of division of labour can reconfigure their production networks more easily to accommodate for smaller environmental changes such as demand fluctuations or changes in consumer preferences – a notion underlying much of the discussion on the competitive advantage of industrial districts over integrated firms in fashion-driven markets. As a result, while there is an optimum degree of division of labour in the case of bifurcation and perturbation events, the fact that this degree is identical might be a model-inherent property related to the fitness landscape.

Regarding the relative performance of different co-ordination mechanisms, the collective regime again performs best both in terms of average cluster fitness as well as with respect to the stability of adaptation processes (see Fig. 8.9.). The alliance scenario yields very similar performance values, albeit at the price of greater fluctuations during the adaptation process (as reflected by greater standard deviations). The individualistic case is the worst co-ordination mechanism leading to low performance and high instability in adaptation. For small parameter values, this picture is the same for the leader firm scenario, however, with C>4.0, its performance improves markedly and once C=8.5, the leader firm scenario catches up with the alliance one thanks to its greater stability.

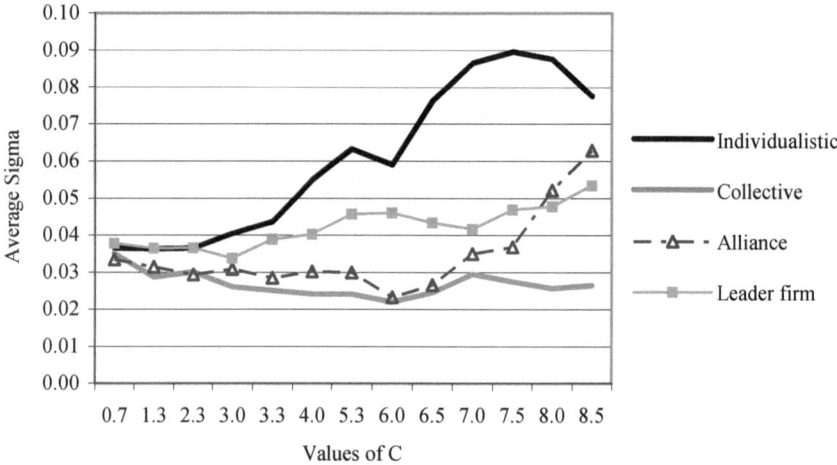

Fig. 8.9. Division of labour, co-ordination and stability in adaptation

Similarly to the case of bifurcations, *adaptive performance of clusters following perturbation events decreases with fitness landscape complexity* (proposition 1). The difference in performance of the co-ordination mechanisms again increases as interdependence between agents grows, i.e. *co-ordination begins to matter more as the role of cross-agent externalities (C) increases* (proposition 2). With respect to their relative performance, those co-ordination mechanisms accounting for a greater share of C externalities (collective, alliance) perform better than individualistic or leader firm cases, i.e. *systems accounting for more C interdependencies exhibit better and more stable adaptation processes* (proposition 3). Finally, although results beyond C=8.5 (EvenK=17) are unavailable, the trends in average fitness and sigma indicate that the leader firm scenario might come to out-perform the alliance one in very interdependent clusters suggesting that the ideas developed in Sect. 8.1 regarding the number, optimality and spread of agent modifications as well a the notion of increased stability through alignment of subordinates' goals and ignorance of some C externalities also hold for adaptation to perturbation events.

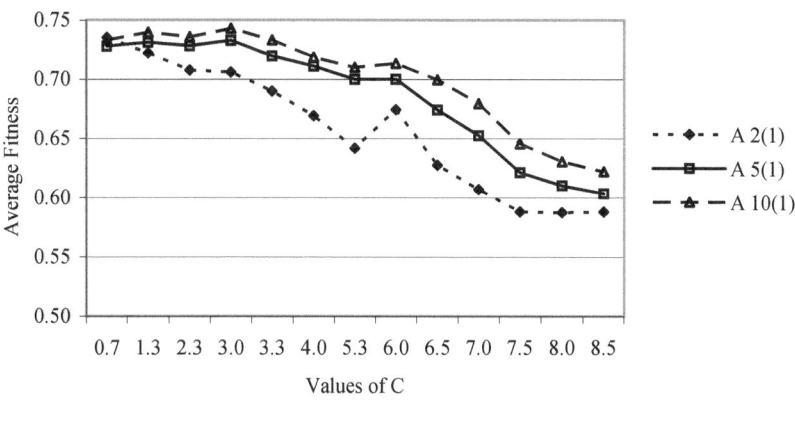

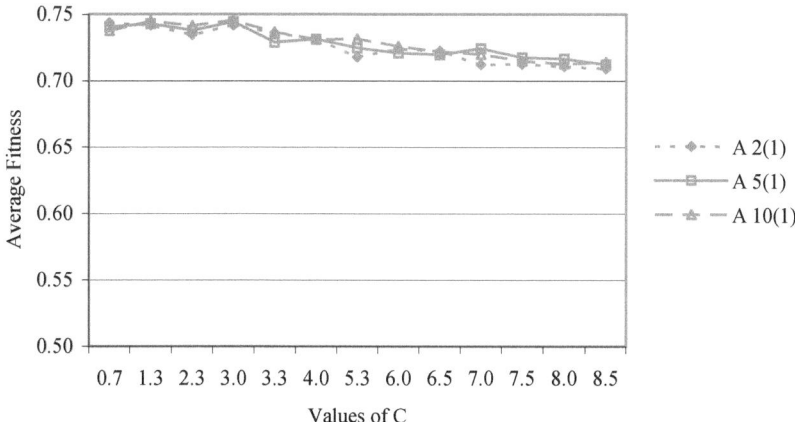

Fig. 8.10. Agent numbers and adaptation: Individualistic and collective

As might be expected from this, the impact of different agent numbers mirrors the case of bifurcations: Greater numbers of agents increase adaptive performance but to differing extents. In co-ordination mechanisms generating a greater number of modifications at the level of their agents, a greater number of actors in each group shifts experimentation from the system to the group level. As a result, better configurations emerge from the bidding process in each group and increase adaptive performance. This effect works more strongly in individualistic and leader firm clusters as compared to alliance or collective ones (Figs. 8.10. and 8.11.).

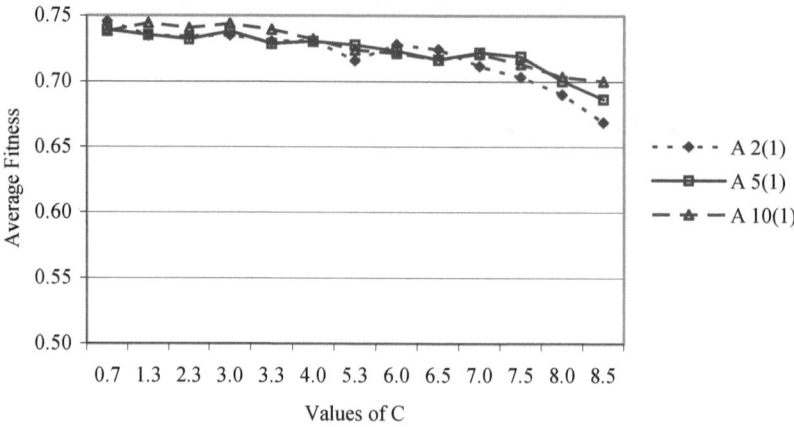

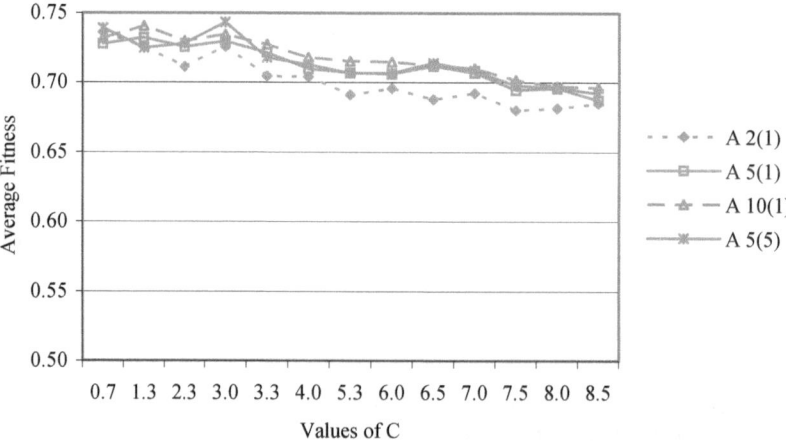

Fig. 8.11. Agent numbers and adaptation: Alliance and leader firm clusters

Summing up, the results obtained before with respect to the role of division of labour and co-ordination also hold in the case of perturbations. Both factors matter for cluster adaptability by shaping the number, system-level optimality and spread of selected agent (group) modifications. Clusters with more collective agent orientations thereby perform better in adaptation to any kind of change event (on average) while the stability effect of a leading cluster agent only materialises under very specific conditions (see Sect. 8.1).

8.3 The tragedy of collectivists

The previous two sections have found that clusters characterised by agents with collective mindsets perform best in adapting to bifurcation as well as perturbation events. Clusters populated by individualistic (egoistic) actors faced great difficulties, especially as cross-agent externalities increased. Alliance and leader firm scenarios positioned themselves in between these two extremes with the former out-performing the latter except for the case of very interdependent clusters with a central leading firm. While the model developed here provides no means for changes in the cluster co-ordination mechanism, their relative performance could act as one explanation for a development away from or towards specific co-ordination mechanisms. Based on the simulation results, any cluster attempting to adapt to changes in its environment would then be poised to move towards the collective mode of co-ordination.

In the empirical literature, however, a trend has been observed shifting co-ordination structures away from egalitarian networks towards the two intermediate modes modelled here (alliance, leader firm). This development has been found in various Italian Industrial Districts (IIDs) as changes in their environment occurred (Boschma and Lambooy 2002; Brusco 1986; Cainelli and Zoboli 2004b; Lombardi 2003; Maggioni 2004). With the results obtained so far from the model, this trend could be justified if agents in IIDs had behaved individualistically beforehand. Moving towards a leader firm or especially an alliance scenario from the individualistic mode of co-ordination almost always increases adaptive performance for the cluster. When investigating the ideal-typical industrial district more closely, however, doubts emerge regarding their correspondence to the individualistic case. While IIDs were composed of networks of egalitarian firms, the strongly emphasised adherence of district firms to an informal local culture could be viewed as evidence that these districts resembled more the collective than the individualistic case (see also Sect. 3.2.1).

While the real world agents in IIDs were probably not identical to the stylised 'Musketeers' represented in the model, their conformity to local rules of the game implies that they had to make their choices not only based on what was best for themselves in any given situation. In a sense, these actors therefore did not behave individualistically. When looking in more detail into some prominent ideas in the literature, considerations like community orientation and entrepreneurs trying to contribute to local development come into play. As a result, one could argue that IIDs were closer to the collective mode of co-ordination. In this case, their shift towards alliance and leader firm structures is somewhat at odds with the results found in the model: If egalitarian networks of collectively oriented agents perform best in adapting to a changing environment, why would the empirical reality show a trend towards alliance or leader firm structures?

A first possible explanation of this phenomenon concerns differences in optimality of co-ordination mechanisms (governance structures) between levels of analysis. In other words, a shift in co-ordination mode could be explained if the collective case worked well for the cluster as a whole but not necessarily for individual firm groups. In such a setting, agents in a group would have an incentive move away from the collective mode to one that increases their individual fitness. The results presented in tables 8.1. and 8.2. however indicate that this intuition is insufficient to explain all shifts in inter-agent co-ordination. Being the leader firm increases fitness for all parameters as is apparent from the comparison of performance for group one in the collective and leader case (see italics in tables 8.1. and 8.2.). As can be obtained from tables A6.4. and A6.7. in appendix 6, the same holds for groups two to four. For some parameter values (see underlined values in both tables), certain groups also exhibit a higher fitness in the case of alliances between groups 1+2 and 3+4. However, only two parameter constellations exist in which the alliance scenario performs better than the collective case for *both* partner groups (see * in table 8.2.). Instead, the collective case maximises group fitness in most instances. What these results highlight is that the only consistent explanation for shifts in mechanism attributable to agent group incentives is the case of the leader firm, i.e. all agent groups in the cluster would benefit if they managed to become the dominant cluster agent.

Analysing group fitness under different co-ordination mechanisms thereby only explains shifts in cluster co-ordination from collective to leader firm cases: If all agents have an incentive to become the leading firm, they will use any opportunity to do so. This has repercussions on the adaptability of the cluster as a whole as was discussed in the previous sections: The gains in fitness for the lead agent (group) come at the expense of adaptiveness in the cluster as a whole. This loss in fitness is especially pronounced for those agent groups that are very strongly connected to the leader firm (see also tables A6.4. and A6.7. in appendix 6). As interdependence between groups one, two and three increases, having the dominant firm in the first group leads to very low performances of that co-ordination mechanism for group two and three. A shift towards an alliance scenario cannot be explained by this analysis since there are too few parameter constellations in which the alliance scenario performs better than the collective case for the agent groups involved. Put differently, there are almost no parameter settings in which the alliance scenario performs better than the collective one for groups one and two or three and four respectively.

Table 8.1. Governance structure and average group fitness (bifurcation)

C	Individualistic				Collective			
	Group 1	Group 2	Group 3	Group 4	Group 1	Group 2	Group 3	Group 4
0.67	0.74021	0.72711	0.72341	0.72405	0.75065	0.73130	0.73066	0.73480
1.30	0.73321	0.73146	0.72241	0.73605	0.74574	0.73890	0.72549	0.74982
2.33	0.75869	0.73424	0.71778	0.72164	0.75424	0.73891	0.73037	0.73332
3.00	0.72625	0.70140	0.72601	0.77043	0.74098	0.72820	0.73268	0.77036
3.33	0.74243	0.71363	0.69287	0.74319	0.74657	0.72040	0.73156	0.73269
4.33	0.70960	0.69163	0.71242	0.70906	0.73351	0.72922	0.70919	0.73804
5.33	0.69894	0.68292	0.69303	0.68972	0.72490	0.71338	0.71393	0.72402
6.00	0.69998	0.69914	0.69194	0.69213	0.72029	0.71858	0.71866	0.71856
6.50	0.66022	0.65744	0.67323	0.68141	0.70661	0.71302	0.72017	0.72162
7.00	0.62246	0.62575	0.65667	0.68476	0.70552	0.70590	0.72192	0.72221
7.50	0.59270	0.59284	0.63012	0.67618	0.68916	0.69571	0.70496	0.74488
8.00	0.57094	0.57144	0.60452	0.68277	0.69103	0.69169	0.70492	0.73155
8.50	0.56044	0.55942	0.58263	0.72636	0.67396	0.68799	0.71015	0.76077
9.00	0.54747	0.54743	0.54709	0.77426	0.69037	0.68189	0.68045	0.77428
10.50	0.53904	0.53950	0.53913	0.71673	0.68574	0.68600	0.67746	0.74776
12.00	0.52996	0.53037	0.53060	0.65417	0.66943	0.68094	0.67271	0.73022
13.50	0.52528	0.52425	0.52458	0.60428	0.66407	0.67500	0.67105	0.70176
15.00	0.51791	0.51807	0.51898	0.56390	0.67794	0.67685	0.66932	0.66838
16.50	0.51502	0.51510	0.51583	0.52647	0.67042	0.65476	0.66795	0.66912
18.00	0.51343	0.51367	0.51355	0.51269	0.66550	0.67176	0.66407	0.66333

C	Alliance (1+2, 3+4)				Leader firm (in group 1)			
	Group 1	Group 2	Group 3	Group 4	Group 1	Group 2	Group 3	Group 4
0.67	0.74523	_0.73523_	_0.73717_	0.73127	_0.75557_	0.70339	0.72173	0.72380
1.30	0.74850	0.73223	0.71915	0.75068	_0.75467_	0.70688	0.72051	0.73730
2.33	0.75214	_0.75074_	0.70799	0.73484	_0.75244_	0.70850	0.71517	0.71828
3.00	0.71935	0.72026	_0.74210_	0.77035	_0.76824_	0.64808	0.70664	0.74475
3.33	0.74438	_0.72758_	0.71503	_0.74525_	_0.76341_	0.62881	0.74073	0.77076
4.33	0.72914	0.70469	0.72196	0.73191	_0.76819_	0.62604	0.73328	0.71198
5.33	0.72171	0.70427	_0.72340_	0.71858	_0.77092_	0.61780	0.71093	0.69709
6.00	0.71585	_0.73105_	0.71433	0.71308	_0.77308_	0.62873	0.69138	0.69286
6.50	0.70652	0.71297	0.70194	_0.72726_	_0.78513_	0.62785	0.67846	0.70398
7.00	0.70529	0.70245	0.70612	_0.72924_	_0.76673_	0.63273	0.65661	0.72184
7.50	_0.69218_	0.69300	_0.71219_	0.73212	_0.75930_	0.62928	0.63677	0.73612
8.00	0.66969	0.67354	0.70363	0.72809	_0.75124_	0.61840	0.62738	0.73681
8.50	0.64706	0.64790	0.68606	_0.76181_	_0.74617_	0.61687	0.61493	0.75696
9.00	0.62163	0.61432	0.63809	0.77426	_0.72174_	0.61286	0.61234	0.77386
10.50	0.61758	0.61934	0.62941	0.72888	_0.71422_	0.60917	0.60780	0.71955
12.00	0.60833	0.61544	0.62134	0.68629	_0.69955_	0.59995	0.59749	0.67106
13.50	0.60401	0.60286	0.60614	0.64899	_0.67929_	0.59022	0.59091	0.63615
15.00	0.58969	0.58810	0.58884	0.61568	_0.67191_	0.57849	0.58297	0.60703
16.50	0.59260	0.59398	0.59048	0.59360	_0.66405_	0.57510	0.58003	0.57332
18.00	0.58022	0.57709	0.58061	0.57754	_0.63426_	0.56735	0.56055	0.56452

Table 8.2. Governance structure and average group fitness (perturbation)

	Individualistic				Collective			
C	Group 1	Group 2	Group 3	Group 4	Group 1	Group 2	Group 3	Group 4
0.67	0.73260	0.72405	0.72638	0.72896	0.74116	0.73787	0.74334	0.73909
1.30	0.74478	0.71520	0.72891	0.73667	0.76510	0.71915	0.73425	0.75279
2.33	0.76350	0.72588	0.71509	0.70958	0.76069	0.72125	0.75740	0.71124
3.00	0.73746	0.70769	0.72627	0.76018	0.74697	0.73755	0.73161	0.76034
3.33	0.73468	0.70018	0.70641	0.73768	0.73003	0.73823	0.72404	0.72308
4.33	0.71987	0.70128	0.71098	0.71198	0.74082	0.72419	0.70858	0.75093
5.33	0.70343	0.69292	0.70395	0.70020	0.73201	0.72381	0.73837	0.70502
6.00	0.69244	0.69803	0.70477	0.70424	0.72450	0.71857	0.72861	0.71170
6.50	0.66335	0.66760	0.68153	0.68398	0.70299	0.72340	0.72857	0.72355
7.00	0.62990	0.63066	0.66505	0.68425	0.70947	0.72941	0.73246	0.72546
7.50	0.59312	0.59531	0.62838	0.66819	0.71028	0.72530	0.71918	0.71484
8.00	0.57506	0.57556	0.60683	0.68326	0.70510	0.69627	0.71940	0.74504
8.50	0.55471	0.55615	0.57738	0.72591	0.68631	0.69272	0.70135	0.76789

	Alliance (1+2, 3+4)				Leader firm (in group 1)			
C	Group 1	Group 2	Group 3	Group 4	Group 1	Group 2	Group 3	Group 4
0.67	<u>0.74230</u>	0.73611	0.74283	0.73571	*0.75437*	0.69146	0.72751	0.73820
1.30	0.76101	0.70090	0.72843	0.75115	*0.76691*	0.69635	0.73072	0.73564
2.33	0.74651	<u>0.73903</u>	0.72003	<u>0.72283</u>	*0.77150*	0.69986	0.72211	0.70911
3.00	0.73427	0.71515	<u>0.74226</u>	0.76023	*0.79331*	0.63369	0.73143	0.76026
3.33	0.72940	0.72466	<u>0.72778*</u>	<u>0.73312*</u>	*0.77015*	0.66288	0.71403	0.73870
4.33	0.73822	0.71852	<u>0.71807</u>	0.74675	*0.76852*	0.62057	0.72716	0.72296
5.33	0.73870	0.72129	0.72821	<u>0.72265</u>	*0.78296*	0.61863	0.72014	0.70604
6.00	0.71451	<u>0.73123</u>	0.70226	<u>0.74455</u>	*0.77971*	0.64389	0.70157	0.69899
6.50	0.71556	0.71943	0.70356	<u>0.72770</u>	*0.78815*	0.65001	0.68500	0.72412
7.00	0.71282	0.71091	0.73714	<u>0.72646</u>	*0.78594*	0.63884	0.66112	0.74183
7.50	0.70621	0.70675	<u>0.72580*</u>	<u>0.73716*</u>	*0.77151*	0.64282	0.62794	0.73643
8.00	0.67703	0.67533	0.69619	<u>0.75313</u>	*0.77650*	0.63831	0.62594	0.74198
8.50	0.64991	0.65140	0.68912	0.75576	*0.75974*	0.61337	0.61492	0.76084

From the analysis conducted so far, the collective case should be the most promi-nent mode of co-ordination in clusters since it maximises fitness for most agent groups. While all agents have an incentive to aim at becoming the leading firm (group), this is not possible for all of them simultaneously. In addition, taking the fitness losses by subordinate groups and at the system level into account would indicate that for the majority of cluster agents, the collective mode is preferable. Moreover, alliance clusters could not be found to out-perform the collective case with respect to the fitness of their involved groups (at least not simultaneously for the allied groups). From this perspective, a move from collective to alliance clus-ters can not be explained.

If the collective mode of co-ordination maximises both cluster and agent group fitness for most parameters and most agent groups, it follows that rational and self-interested cluster agents should adopt this form of co-ordination. As was men-tioned before, this is at odds with empirical observations where the collective mode of co-ordination was found to be unstable in a changing environment. What else could then explain a shift in cluster organisation away from collective towards alliance or leader firm structures? The answer to this question lies with two possible aspects. First, there was an issue raised with respect to the speed of adaptation in the collective case (see Sect. 7.2.2). The collective case would maximise cluster fitness but at the same time, agents would be relatively slow to find improvements in their configuration if evaluation was based on N=24 elements. Instead, the individualistic mode lead to faster improvements in the configurations of individual agents, which were however suboptimal from a cluster perspective. Depending on the extent of interdependence between agents (C), this implied a greater number of changes at the agent level with potentially negative repercussions between agent groups. Nonetheless, the fact that individualistic behaviour leads to a faster encounter of what seems like improvements in the agent's configuration[121] might provide an incentive for agents to abandon the tiresome and long-term collective orientation in favour of an individualistic one.

Whether or not the collective mode of co-ordination can be destabilised by individualistic behaviour depends on the mechanisms supporting agent orientations aimed at improving the situation (fitness) for the cluster as a whole. Agent orientations in clusters were influenced by the nature of the local culture. The latter in turn evolved (see Chaps. 3.1.2 and 4.1) to provide a shared understanding of acceptable business practice to avoid – among other things – a tragedy of the commons with respect to the activities generating agglomeration economies. Agent understanding and corresponding behaviour (adherence to the local culture) was in turn supported by different, more or less 'institutional' mechanisms: Informal local rules of the game, credible threats of using formal institutions (legislation) or differential distributions of power between cluster agents (alliances, dominant actors). In networks with even distributions of power, the enforcement of collective orientations is thus based on some form of collective action excluding defecting agents or on the use of formal institutions against them (Holländer 1990; Caeldries 1996).

While these support mechanisms can be expected to work and gear agent orientations and activities to conform with the rules of the local culture in situations of stability (thanks to the superior opportunities of observing and judging the behaviour of spatially close agents) this may not be the case in the time following an external bifurcation or perturbation event. In times of stability, agents in clusters were argued to develop their mode of co-ordination in a bid to foster positive and limit negative activities between them. While the economic environment is stable, any deviation from these rules of the game could be discovered by its adverse effects on other agents and the cluster as a whole. The very nature of environmental change events, however, implies that this evaluation mechanism loses some of its accuracy. As the competitive environment changes to reduce the fitness of the cluster, individual agents could defect from the local rules of the game (i.e. change

---

[121] After taking the C interdependencies, the actual fitness of the new agent group configuration may be much lower than the expected one since the co-ordination mechanism does not internalise any cross-agent externalities, thereby increasing the number of 'mistakes' that agents make.

their orientation and activity from collective to individualistic) to increase their more immediate benefits (i.e. to find apparent improvements faster). In order to avoid punishment by other local agents, they could then ascribe the effects of their defection on other cluster agents to the change in the greater environment.

To test this intuition on the instability of collective orientations, the simulation model was extended to include the effects of egoistic agent and group behaviour in otherwise collectively oriented clusters. Initially, simulations were run with an increasing number of egoistic agent groups (ranging from one to three out of four) in the cluster. The comparison of average group fitness in these mixed clusters with that found for the individualistic and collective case shows a strong *prisoner's dilemma payoff structure*. In terms of average fitness, any agent group has an incentive to act egoistically as being the one egoistic group in an otherwise collective cluster maximises group fitness. At the same time, any group suffers worst in terms of fitness if it is exposed to individualistic behaviour by its most closely connected neighbour. Having an egoistic neighbour (defined as the group with most C externalities) leads to a group fitness value below that obtained in the individualistic case. Any agent group thus benefits most by being the only egoist in the cluster. It suffers more than in the individualistic case if *only* its neighbour group behaves egoistically. Of course, results in terms of fitness were even worse when the group was the last collectivist remaining in the cluster. In terms of group fitness, the following ranking emerges for all parameter settings and all agent groups (see also Fig. 8.14.): *Solely egoistic>Collective>Individualistic>Egoistic neighbour*. As a result, in the presence of any C externalities between groups, egoistic behaviour will spread in the cluster once one group starts behaving egoistically as egoistic groups 'infect' their respective neighbours.

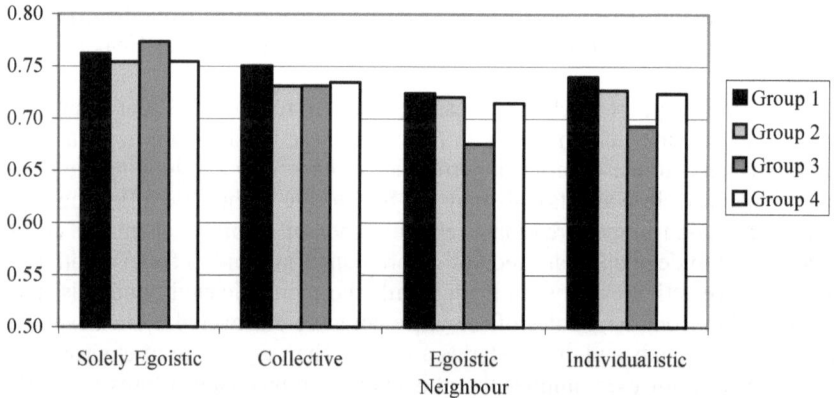

Fig. 8.12. Group behaviour and fitness – Prisoner's dilemma revisited[122]

A second step in investigating the stability of the collective mode of co-ordination then involved introducing egoistic agents into each group to see at what point the

---

[122] For Pm=EvenK=4. Full results can be obtained in tables A6.8. and A6.9.

presence of individualistic agents would tip group behaviour from collective to individualistic. The average fitness for groups with different numbers of egoistic agents was again compared to that found with individualistic and collective clusters. The results are however not systemic in the sense that there is no single number of egoistic agents that tips group behaviour into either direction. In some instances, the aforementioned prisoner's dilemma already emerges with one or two egoistic agents per group whereas in other cases it does not emerge unless the entire group behaves egoistically. This aspect can be attributed to the nature of the bidding process at the group level. As agents are chosen to represent their group based on expected cluster fitness, in groups with egoistic and collective agents, neither type of agent will dominate group behaviour. While egoistic agents (as in the individualistic mode of co-ordination) generate a greater number of modifications, the latter are bound to be suboptimal from a system perspective (see also Sect. 8.1). Collectively oriented agents in turn generate fewer yet better modifications. As a consequence, having both types of agents bid to be the representative at the group level means that both will get chosen with different probabilities. The behaviour of such mixed groups is thereby neither egoistic nor collective which leads to a limitation of its adverse effects on the neighbouring group.

For prisoner's dilemma constellations to materialise and destabilise the collective mode of co-ordination, it is thus necessary that an entire agent group behaves egoistically. All else held equal, clusters with less agents per group are more at risk of such an invasion of egoists. The resulting spread of egoistic behaviour in the cluster then implies that they move from the collective to the individualistic mode of co-ordination as agents increasingly select strategies based on egoistic motives related to their own fitness. This has very adverse repercussions on the adaptability of agent groups and the cluster as a whole. The trend of real world districts to move towards alliance or leader firm structures could then be explained as a way to restrict egoistic behaviour in the face of change events in the environment and thus avoid a fall from the best to the worst possible mode of co-ordination in the model.

8.4 Summary

The simulation model has found that adaptive performance and stability of adaptation processes in clusters were related to the extent of division of labour as well as the mode of co-ordination. The role of both factors in cluster adaptation emerged due to their impact on the number, system-level optimality and spread of selected agent modifications. By determining the structure of the fitness landscape (K, C) the degree of division of labour plays a role for the number and spread of agent modifications. High degrees of interdependence within an agent's subset of the fitness landscape imply that search activity will lock-in to a local optimum more early thereby reducing the number of modifications executed by individual agents. This stabilises cluster adaptation. However, due to specificities of the LSD programme, these high degrees of intra-agent interdependence (K) come together with increases in inter-agent linkages (C). The latter mean that even small changes in one agent's configuration can spread very far in the cluster, deforming the landscape of many other actors. This in turn induces further modifications in the latter's configuration, increasing the instability of adaptation processes. For intermediate degrees of inter-agent linkages (C=3.0), adaptive performance is maximised, indicating an optimum extent of division of labour.

Given a specific degree of division of labour (landscape structure), the mode of co-ordination becomes key to adaptive performance due to its influence on the number and system-level optimality of agents' modifications. In clusters with more collectively oriented agents, fewer improvements of agent configurations are selected as the likelihood of finding better configurations in a given fitness landscape subset decreases with the size of the evaluation landscape. As a consequence, adaptation in clusters with more collective agent mindsets is more stable as individual agents and thereby agent groups change less frequently in configuration. This limits their disturbance on the fitness landscape of other agents. Moreover, clusters with more collective agent orientations also generate better agent modifications from the system perspective by internalising a greater share of cross-agent externalities (C) into the decisions of the former. As a result, for most simulations the collective cluster ranks above alliance, leader firm and individualistic ones. Leader firms only benefit cluster adaptation in comparison to the alliance case if the cluster is highly interdependent and if the leader firm occupies a central position with respect to inter-agent externalities.

The shift in mode of co-ordination (from collective to alliance or leader firm) observed for many Italian Industrial Districts can be explained by two different aspects. In terms of group fitness, every cluster group has an incentive to become the leader as this mode of co-ordination maximises any group's fitness. This cannot be observed for alliance constellations, i.e. there are too few parameter settings in which the fitness of the groups involved in an alliance exceeds that obtained in collective clusters. Second, the collective mode is found to be unstable against the emergence of egoistic behaviour following a change event. In this context, a move towards alliance or leader firm structures can be seen as a strategy to prevent the emergence and spread of egoistic behaviour in clusters. Once established within any cluster group, egoism would diffuse within the cluster thanks to a prisoner's dilemma payoff structure: While any group's fitness benefits most if it is the only egoistic force in the cluster, it suffers most if only the neighbour group behaves egoistically. The level of fitness in this constellation is even lower than for the individualistic case where all agent groups behave egoistically. This implies that any agent group affected by an egoistic neighbour has nothing to lose if it starts behaving egoistically as well. Without intervention (e.g. by joint action of all cluster agents or by an outside actor), this payoff structure fosters a shift from collective to individualistic modes of co-ordination with severe repercussions on cluster and group fitness. Moving towards alliance or leader firm structures depending on the possibilities in each real world case thus constitutes the best possible option for the cluster given the instability of the collective mode of co-ordination. As a result, the developments found in many IIDs may well be more desirable from the district perspective than has often been acknowledged.

9 Model contribution, limitation and avenues for future research

Throughout the development of virtually any kind of economic system, events can occur which so change the conditions for its success or failure that substantial change in the system's current mode of operation has to occur. Just like one will never be able to derive a railroad from adding mail coaches after one another (Schumpeter 1934a, p. 64), small changes in business practice are not likely to help cluster agents survive this kind of event. In such a situation, however, the advantages to being in a cluster can turn into liabilities. This is attributable to the existence of cluster level features (agglomeration externalities, local culture), that are fixed within the short-term following a change event. They imply two things:

- First, the success of agent activities hinges on the actions of other and
- second, not all possible alterations of agent's business operations may be feasible within the context of a given local culture.

In addition, the cluster as the repository of these architectural factors that influence and constrain agent behaviour lacks any central decision-making authority able to co-ordinate local agents for optimal collective results. However, as was found in the study conducted here, some kinds of cluster architectures are more conducive in steering agent-driven adaptation to good (individual and collective) results. This section starts discussing the implications of these findings (9.1) by summarising the causalities found in the modelling and simulation exercise regarding the role of division of labour and co-ordination for cluster adaptation (9.1.1). It then proceeds to outline in how far these causalities can explain the observable empirical realities regarding the shifts in mode of co-ordination in Italian Industrial Districts and the relative performance of Silicon Valley and Boston 128 at the advent of the microcomputer (9.1.2). Sect. 9.2 then discusses the contributions of the model to the existing literature from an empirical as well as a theoretical perspective. It finds that the study adds to a more dynamic perspective in Marshallian cluster concepts by showing that technological externalities and non-linear interdependence are reconcilable with general dynamic models when adopting a complexity perspective. Nevertheless, the model faces a number of limitations (9.3) with respect to agent dynamics, the inclusion explanatory variables as well as the use of methodology, which all invite a number of avenues for future research (9.4). The latter could involve both a testing of the model in empirical settings as well as model extensions. Sect. 9.5 concludes.

9.1 Implications of findings

One of the goals of the study involved finding causalities and explanations for the observable empirical trends describing an influence of the extent of division of labour and the mode of co-ordination on cluster adaptation. Through use of a general theoretic model and simulations, it has become possible to isolate the influence of both factors. This was achieved by investigating and comparing adaptation in clusters differing only according to their *given* division of labour and mode of co-ordination. With these causalities in mind, an explanation of stylised empirical phenomena becomes possible.

9.1.1 Division of labour and co-ordination: The causalities

Regarding the nature and direction of causalities it is found that division of labour and co-ordination matter for cluster adaptation in the model because they influence the ability of the co-evolving N/K(C) system cluster to find better configurations after changes in its fitness landscape. This ability is shaped by the number, system-level optimality and spread of modifications selected by different cluster agents and their groups.

By determining the degree of interdependence within and between agents, the extent of division of labour impacts on the number and spread of agent (group) modifications:

- High levels of intra-agent interdependence (K) imply a more rugged fitness landscape subset for every agent. As modifications in one or more elements increasingly produce changes in the fitness value of others, very interdependent subsystems exhibit a variety of configurations with similar or identical fitness values. Some of these optima may be the global, i.e. the best attainable configurations in the landscape subset, others may constitute local optima, i.e. configurations that are best *given* the initial configuration and search process adopted by the agent. More rugged fitness landscapes at the level of agents therefore imply that their search and modification activity settles in to a (local) optimum more quickly than for lower K values. As a result, agents and agent groups generate fewer modifications.
- The number of modifications generated by agent groups then plays a role for system adaptability. Thanks to their ability to spread in the system (due to inter-group externalities), modifications of one agent group impact and deform the fitness landscape subset of others. The greater these C externalities, the greater the mutual disturbance in agent and group adaptation, which reduces the adaptive performance of the system as a whole.

Due to specificities of the LSD programme, increases in intra-agent externalities came alongside more inter-agent ones, resulting in a trade-off with respect to the role of division of labour. While high values of K externalities would stabilise agent and thereby system adaptation, these parameter settings also lead to higher degrees of C interdependence. The latter meant that even small modifications in an agent group spread far within the system, thus offsetting the stability effect of higher K externalities. What emerged from the simulations was an *optimum degree of division of labour*. For limited landscape complexity, the literature on N/K has revealed that average fitness of (local) optima increases as compared to simpler landscapes. This is attributable to a greater number of random draws for the (conditional) fitness values of individual elements. Beyond this optimum complexity level, however, the aforementioned problems of optimising very interdependent co-evolving systems decrease adaptive performance. For a cluster of N=24 elements controlled by four agent groups with n=6 elements each, the optimum degree of landscape complexity materialised for an intra-agent interdependence of K=4.3 and an inter-agent one of C=3.0 (EvenK=9).

Depending on the degree of division of labour in the cluster, agent groups controlling one subset of all value chain activities walk on landscapes that are deformed by the activities of other agents. Individual agents cannot influence others' activities to the extent that a central decision-making authority (e.g. in integrated firms) could. The only way to control for some of the effects of inter-agent interdependence resides with a mode of co-ordination between local actors. Building on the existing literature on clusters, this study assumes four ideal-typical mechanisms that may have evolved throughout the cluster's previous history: Individualistic, collective, alliance and leader firm constellations. The mode of co-ordination, once established exerts an influence on the goals underlying agent strategies: Depending on the local rules of the game that have been agreed upon (the local culture), different kinds of business practice are viewed as (un)desirable. The modes of co-ordination investigated here differ in their degree of collectiveness, i.e. the extent to which agents take the effect of their activities on other cluster actors into account when choosing a strategy. The more collective these orientations, the fewer changes in agent strategy will occur since the likelihood of finding improvements on the agent's fitness landscape subset decreases with the size of the evaluation landscape, i.e. the extent to which the fitness of other agents enters individual decision-making. By influencing agent strategy selection, the mode of co-ordination therefore impacts on the number and system-level optimality of modifications:

-   While agents in clusters with collective mindsets take longer to find improvements in the configuration of their elements, their proposed modifications will be better from the perspective of the entire system. As a result, clusters with more collective agent mindsets (collective, alliance) outperform more individualistic ones (leader firm, individualistic) both in terms of average fitness as well as fluctuations in system fitness throughout their adaptation. This is attributable to the lower number of better modifications executed by agents and their groups, which improves and stabilises adaptation in co-evolving systems.

-   For very specific fitness landscape conditions, the leader firm case can outperform its closest competitor regarding agent dynamics (alliance). Initially, the lack of interdependence of some agent groups with the leader firm and the latter's individualistic behaviour imply strong mutual disturbance in adaptation as agents' selection mechanisms do not reflect the actual distribution of C externalities. For the stability and performance benefits of having a leading firm to materialise, the dominant agent has to be central to the cluster (generate and receive externalities from all other agent groups) and relatively inactive (high K values for the dominant firm). In this case, the alignment of subordinate agent goals trying to improve the leader firm's fitness alongside their own stabilises system adaptation. Moreover, with high K values for the leader firm, the activities of subordinates are left relatively undisturbed.

The number of agents in each cluster group also plays a role for system adaptability. With greater agent numbers, experimentation with group configurations shifts from the system to the group level. Only very good configurations are able to win the bidding process in large agent groups, thereby decreasing the number of modifications for each agent group and increasing their optimality from the system's perspective (thanks to the nature of the bidding process, which is based on expected cluster fitness; Sect. 7.1.4). This effect of agent numbers materialises more strongly for co-ordination mechanisms where agents generate a greater number of modifications, i.e. it works more strongly to increase adaptive performance for individualistic and leader firm cases while its role for alliance and especially collective clusters is more limited.

These findings on the performance effect of division of labour, co-ordination and agent numbers apply for adaptation to bifurcations and perturbations. Nonetheless, adapting to perturbation events will come easier to a cluster as parts of the system may be unaffected by changes in the fitness landscape. Overall cluster (system) performance can thereby benefit from these agent groups retaining any high fitness value obtained prior to the change.

9.1.2 Explaining empirical trends: Districts and Silicon Valley

With respect to offering an explanation for the stylised empirical phenomena,[123] the aforementioned causalities alone are insufficient. On the one hand, empirical research in Italian Industrial Districts (IIDs) shows a shift in their co-ordination mechanism towards alliance or leader-firm structures. From the insights obtained on the role of co-ordination, any shift aimed at increasing adaptive performance of the cluster would however have to be directed at the collective mode. On the other hand, the Silicon Valley-Boston 128 comparison highlights an adaptiveness advantage of networked small firm clusters over those composed of large, integrated enterprises. Regarding the results found for the role of division of labour alone, this would only be possible if the degree of decentralisation of productive activity had not become too extreme.

---

[123] Of course, other industry- or country-specific causalities may underlie the findings of the Silicon Valley – Boston 128 comparison or the Italian District experience (see also Sect. 9.3).

To explain the first phenomenon – the shift in co-ordination in IIDs – two separate analyses were conducted. A first step involved a comparison of adaptive performance between the system and the agent group level. It is not surprising that collective clusters composed of agents gearing their strategy selection at improvement of the entire system should perform best in adaptation for the cluster level. This aspect was not that obvious at the level of agent groups. A first possible explanation for shifts in co-ordination mechanism would therefore lie with contradictions in optimality between different levels of analysis. If alliance or leader firm scenarios worked better for the fitness of agent groups, these groups would have an incentive to move towards these co-ordination mechanisms, even if this entailed a loss in adaptive performance for the entire cluster. The comparison of system and group fitness under different co-ordination mechanisms however revealed that the collective mode also maximised group performance in most cases. The exception was the case in which a group came to be the leader firm. As a consequence, any agent group had an incentive to become dominant in the cluster. This could explain a shift in co-ordination from collective to leader firm scenarios given the ability of an agent group to come to dominate the cluster. A shift towards alliance structures could not be explained with better agent group performance as there were too few parameter constellations in which alliances increased the fitness of the partners involved (relative to the collective case).

A second possible explanation for the shift towards alliance and leader firm constellations in IIDs was then found to lie with the instability of the agent orientations underlying collective clusters. As was established before, agents wanting to improve system fitness through search and modification activity within their landscape subset took longer to encounter improvements than those gearing search and modification at their own fitness alone. While the improvements discovered by agents with individualistic mindsets may only be *apparent* (depending on the extent of C externalities and the activities by other cluster agents) agents with either type of orientation would have no way of knowing beforehand whether their modifications will lead to a fulfilment of their goals since they cannot fully anticipate the activities of other actors (Allen 1997; Axelrod and Cohen 1999 – see also Chap. 7.1.3). Alongside with the decreased ability of other cluster actors to detect agents departing from the established rules of the game after a change event (see Sect. 8.3), individual agents and agent groups might be tempted to move towards more individualistic mindsets for the sake of finding (apparent) improvements in their configuration more quickly.

Investigating the relative performance of agent groups with individualistic and collective mindsets then confirmed the aforementioned intuition. Group fitness levels exhibited a strong prisoner's dilemma payoff structure where any agent group benefited most from being the sole egoistic (individualistically minded) group in the cluster while suffering worst if its most connected 'neighbour' group behaved individualistically. Comparing average group fitness the following constellation emerged: *Solely egoistic> Collective> Individualistic> Egoistic neighbour*. With any interdependence between agent groups, egoistic behaviour would thus spread in the cluster once one agent group started defecting from the collective mindset. This then led to a shift in the cluster's mode of co-ordination from the collective to an individualistic scenario with very adverse repercussions on system and agent-group fitness values. Moreover, it was found that for this prisoner's dilemma to emerge, all agents in a group had to behave individualistically. Clusters with fewer agents in any of their groups would therefore be more liable to this destabilisation of collective co-ordination through the invasion of egoistic groups.

The move of IIDs towards alliance and leader firm structures can now be explained by two different rationales. First, any agent group in the cluster has an incentive to become the dominant actor as this maximises group fitness. Moreover, in the presence of change events, formerly collective agent orientations could change, thereby destabilising the corresponding mode of co-ordination. Moving towards alliance or leader firm structures can then be a means to avoid a deterioration of the collective cluster into an individualistic one. As a result, the shift in co-ordination form within IIDs may be more desirable from the district's perspective than has usually been acknowledged.

The question of why Silicon Valley's computing industry outperformed that of Boston's Route 128 after the introduction of the microcomputer is a more difficult one. With respect to the role of the extent of division of labour alone, the better performance of the Silicon Valley might be attributed to the fact that its division of labour led to a degree of fitness landscape complexity that was closer to the optimum one (C=3.00) than that of Boston's Route 128. However, the degree to which related activities were conducted within integrated or between independent firms was not the only difference between both clusters. It is entirely possible that they also exhibited different modes of co-ordination. Assuming different modes of co-ordination however has a crucial impact on adaptive performance (see previous section). Moreover, the smaller size of Silicon Valley firms could imply that there were more of them at each stage of the value chain than in the Boston case, which has been found to increase adaptive performance. How the causalities and potential trade-offs between the degree of division of labour, the mode of inter-agent co-ordination as well as the number of agents per group play out for cluster adaptiveness in the case of multiple differences between them will have to be resolved in future work. It can however be said that more collective mindsets and greater agent numbers would speak in favour of more decentralised clusters like the Silicon Valley although the range of landscape parameters in which they can outperform more integrated ones may be limited due to the dominant role of fitness landscape complexity for adaptation (see Sects. 8.1 and 8.2).

9.2 Model contribution

The model developed in this study investigated the role of division of labour and co-ordination mechanisms in clusters for their adaptability to change events. By deriving general causal links between specific factor constellations and the performance of cluster adaptation, it has achieved a number of contributions to the existing literature. With respect to empirical research and policy initiatives, the model has clarified one possible set of causalities underlying the observed trends in adaptation for the Italian Industrial Districts (IIDs) and to some extent the Silicon Valley – Boston 128 comparison. The explanation was based on an analysis of the role of both factors for steering agent-driven adaptation in clusters towards good collective outcomes. By isolating their respective influence within a stylised model of clusters, causalities have been derived in a more conclusive fashion than was possible from case study evidence alone. It has highlighted that the developments in IIDs are more suitable to their performance in a changing environment than has usually been acknowledged. Moreover, a first step towards an explanation of why, how and when decentralised clusters can out-perform more integrated ones in adapting to change was provided. Third and most importantly, the model's generation of causal links can inform future empirical research on factors worth taking into account when investigating cluster adaptation to change. Policy initiatives in turn might benefit from the findings generated here insofar as the model (within limitations; see Sect. 9.3) can derive insight into whether an existing cluster is likely to perform well in the presence of external changes as well as whether trends in the existing industry regarding its division of labour or mode of co-ordination are geared towards helping or hampering agent-driven adaptation.

In addition, the model extends the complexity and N/K literature in two ways. First, it introduces a new application to the framework by understanding clusters as complex, co-evolving N/K(C) systems. This understanding however implies a number of changes to the original model to reflect the nature of clusters as constructs composed of interdependent agent groups that all act upon the activities under their control *in absence* of a central decision-making authority. Adaptation in clusters therefore had to be modelled as a case of decentralised problem solving under cluster-level interdependence. Moreover, the agents in this co-evolving system are able of conscious and deliberate search and modification activity, i.e. they are able to choose their selection mechanism. This also implies that they may be aware of the existence of some degree of system-level interdependence, although the exact effects of the latter are unknown to them. By allowing for more collective mindsets where agents evaluate their activities including elements *outside* their own control, this model then resolves some of the usual trade-off between speed (decomposition size) and performance (system-level optimality) of adaptation in N/K(C) systems. While agents trying to improve the fitness of all ($N=24$) or some ($n=12$) elements through search over the $n=6$ elements under their control will be less likely (and thereby slower) to find improvements in their configuration than agents basing search and selection on an $n=6$ landscape, this trade-off is less pronounced than if agents gear their search at the same landscape size. Put differently, an agent searching on an $n=6$ landscape to improve an $N=24$ element one will be faster in doing so than an agent searching and improving an $N=24$ landscape using the same search process (as has been predominant in the existing N/K literature).

The factors that the study investigates in analysing cluster adaptability (division of labour, mode of co-ordination) are also relevant in other fields of science such as organisation theory (Carley and Lee 1998; Carley and Svoboda 1996; Chang and Harrington 2000, 2004; Ethiraj and Levinthal 2004a; Kollman et al. 2000 or

Marengo and Dosi 2005). However, a key difference between the role of both factors for adaptation in *the case of firms versus the case of clusters* regards the absence of a central decision-making and decision-enforcement authority in the latter. The phenomenon studied here – decentralised problem solving by agents under system-level externalities without central planning or authority – could moreover be imagined to apply for a wider class of (social) systems than the one investigated here. Instances could be systems where performance is shaped by the activities of agents and where actions by one can help or deteriorate the success of actions by another. If furthermore agent activity cannot be fully controlled by a central decision-making authority (principal-agent and incentive problems), these systems approach the case modelled here. It remains to be seen whether the conditions underlying the modelling and simulation exercise here are met by other phenomena and to what extent the results found here hold in the latter as well.

Finally, this study would like to address itself to the proponents of the Marshallian cluster concepts described in the third chapter. Despite all criticism, this research has progressed towards an understanding of the mechanisms and effects of agglomeration externalities. Against this background, it is argued here that a more dynamic orientation on cluster development is needed, not only regarding their emergence but also with respect to *exhaustion and decline*. While having its limitations (see the following section), this study has provided a first step into this direction by showing that a complexity framework is able to reconcile the existence of technological externalities and non-linear interdependencies with a dynamic perspective. While the comparison of dynamics for given cluster architectures done here is still constrained with respect to its ability to address more evolutionary questions, linking cluster and complexity research is believed to provide a very fruitful way towards understanding some of the very pressing issues involving clusters, change and adaptation.

9.3 Model limitations

In investigating whether, how and when cluster agents can adapt to changes in their environment, the model has focussed on a small subset of all possible factors by analysing the role of architectural factors residing at the cluster level. These factors were found to steer agent and agent group activities by shaping the goals underlying them (co-ordination) as well as determining the repercussions between agent activities (division of labour). By limiting the analysis to these factors, important other aspects have been omitted.

A first restriction of the model regards the treatment of cluster agents. While these are argued to shape cluster behaviour, the model lacks important drivers of agent dynamics through its assumption of representative organisations. In consequence, issues like organisation-level inertia and agent heterogeneity with respect to their ability to recognise and accommodate change do not come into play. In addition, learning is currently unaccounted for since the behaviour of cluster agents remains the same regardless of the time they spend adapting. Moreover, the model only investigates two out of three stylised empirical factors argued to affect the adaptability and survival of clusters. The third aspect, internationalisation of cluster actors was left out of the analysis since it does not constitute an architectural, cluster-level property in the strict sense advocated here. Including agent internationalisation through different agent search processes (see Sect. 9.4) would be a first step towards accounting for agent inertia and heterogeneity by endowing internationalised cluster agents with better performing search processes and investigating how this matters for overall adaptive performance.

The explanation provided by the model for the observable empirical trends is one out of many possible causes of the latter. It has been encountered within the stylised world of N/K systems based on a specific perspective on the nature of clusters. Within this understanding, division of labour and co-ordination mattered for cluster adaptation by influencing the number, (system-level) optimality and spread of modifications adopted by each agent group. However, numerous other factors could have generated the observed empirical trends such as general industry developments favouring a concentration of firms (for IIDs) or a different impact of the advent of the microcomputer benefiting Silicon Valley even in absence of any activity by agents in that cluster. It was never the goal of this study to provide the one and only determinant of the observed realities but to test if the model itself could generate an explanation of the observed phenomena (Gell-Mann 2002). In this context, the modelling and simulation exercise has been successful.

Nonetheless, some results provided by the model are methodology-inherent. As an example, the predominant role of the division of labour and thereby the fitness landscape structure for adaptive performance of clusters is attributable to the landscape's role in measuring system performance on the one as well as the general properties of N/K system fitness landscapes on the other hand. Landscape properties also underlie the counter-intuitive result of an identical optimum degree of division of labour in the case of bifurcation *and* perturbation. Moreover, the representation of clusters within the N/K framework adopted here is admittedly stylised and based on a specific understanding of the phenomenon which might well be challenged by different perspectives on what constitutes the nature of clusters.

With respect to model dynamics, the current implementation faces the problem that in order to derive causalities regarding the roles of division of labour and co-ordination, the model takes both as given and proceeds to conduct what could be labelled *comparative dynamics within static architectures*. As a result, there are no opportunities for cluster agents to change their division of labour or their mode of co-ordination. Moreover, the stasis of architecture justified by the model's current short-term horizon does not allow for the effects of selection and emergence at the level of cluster agents. When attempting to provide a more evolutionary picture of clusters, local bifurcations and adaptation, both aspects would have to be accounted for.

Finally, a word of caution has to be advanced with respect to the link between model findings and cluster survival. Two reasons are responsible for this. First, the use of complex system theory implies a breach with deterministic predictions. Contrary to mechanical systems, complex system behaviour is not entirely predictable. Within the N/K framework, the randomness of fitness landscapes alongside the possibility of path dependence and lock-in of agent search processes introduce a portion of luck into the outcomes of agent-driven adaptation. While the results and insights gained correspond to adaptive performance in the majority of cases, they are more to be viewed as tendencies. Put differently, the right division of labour and collective modes of co-ordination mean that agents in such clusters have a higher probability to adapt well to a change event. In individual simulations, the results reported here on the roles of co-ordination and division of labour were not necessarily met every time.[124]

Second, change and adaptation proceed through different phases involving aspects of self-organisation and arbitrage both at the level of individual clusters as well as their host area (see also Chaps. 1 and 4). The model developed here has focussed on one of the two aspects (agent dynamics and architectural factors) shaping the likelihood of success in adaptation by self-organisation in clusters directly following a change event. Since events in the second phase of adaptation will also impact on cluster survival (e.g. regarding the relative competitiveness of the cluster relative to other clusters or integrated firms in the industry), the model findings do not have a one-to-one relationship to cluster survival. Put differently, clusters exhibiting an optimal degree of division of labour and very collective agent mindsets are *more likely* to start out on a positive development trajectory after the change event. This does not mean that they will always survive. The issue of survival or decline will also depend on their performance in the second phase of adaptation, which the model does not address. Both aspects limit the empirical testability of model results as well as their value for policy initiatives.

In consequence, any empirical tests or political recommendations that might be derived from the model results have to take into account that the latter are to be viewed as tendencies rather than predictions. First, the model takes a very ideal-typical idea on the nature of clusters as its point of departure. Second, there is a multitude of interdependencies influencing causal relationships in real-world clusters out of which only some are accounted for here. Third, even within the stylised world of N/K systems, the complexity of clusters implies that systems with inferior architectures might still get lucky in adapting to change. As a result, while the model indicates that certain cluster architectures are more conducive to adaptation, it does not stake the claim that differently configured clusters can not adapt well to change events. However, successful adaptation in this context is more an outcome of luck than of any suitable steering properties at the cluster level.

---

[124] To illustrate the case in point on complexity and prediction, Alfred Marshall used the example of astronomy: *"Thus, having studied the lie of the land and the water all round the British isles, people can calculate beforehand when the tide will probably be at its highest on any day at London Bridge or at Gloucester; and how high it will be there. For, though many forces act upon Jupiter and his satellites, each one of them acts in a definite manner which can be predicted beforehand: but no one knows enough about the weather to be able to say beforehand how it will act."* And: *"The laws of economic are to be compared with the laws of the tides, rather than with the simple and exact law of gravitation. For the actions of men are so various and uncertain [and interdependent], that the best statement of tendencies, which we can make in a science of human conduct, must needs be inexact and faulty [when applied to individual cases]"* (Marshall 1920, p. 32; own emphasis).

9.4 Avenues for future research

The aforementioned model limitations as well as its insights generate a number of avenues for future research. With respect to model insights, a possible strand of investigation would involve their empirical testing. This approach will however find a number of difficulties with respect to controlling for unobserved variables. First, the nature and extent of the change event will matter for cluster adaptability, i.e. one would have to ask whether the cluster faced a (beneficial or adverse) Schumpeterian bifurcation or whether it was challenged by a perturbation event. Second, macro-level factors beyond the cluster itself might impact on its adaptation as for example national legislation or industry-specific factors. Third, a number of variables or properties may differ between agents in different clusters, thereby leading to divergence in adaptive performances for each case. To some extent, these issues could be controlled for by trying to find very similar empirical cases. Comparing adaptive performance in clusters within the same industry or country undergoing the same change event would be a first step, although the lack of explicit agent dynamics in the current model remains problematic. Despite all these concerns, the evidence on Italian Industrial Districts already shows that case studies can point at overreaching trends. It remains to be seen whether the shift in mode of co-ordination is a purely Italian issue or whether other supporting examples can be found.

A second avenue for future research involves model extensions. Among the more immediate ones is the investigation of the trade-offs between the three aspects driving cluster adaptability in the model: The extent of division of labour, the mode of co-ordination and the number of cluster agents in each group. In a bid to revisit the Silicon Valley-Boston 128 comparison, simulations could be run to determine when a decentralised cluster with many agents can out-perform a more integrated one with fewer organisations per group. This exercise can be conducted using suitable assumptions regarding the co-ordination mechanisms in both clusters. One possible line of argument would be that the lower agglomeration externalities between the large integrated firms in Boston 128 allowed for a more individualistic agent mindset while Silicon Valley firms were more collective in their orientations. Moreover, the trade-off between the effects of different determinants of adaptability could be investigated by increasing the frequency of environmental perturbations. This should also shed additional light on the extent of the trade-off between speed and system-level optimality of agent modifications in individualistic and collective clusters.

Beyond this aspect, the model could be extended to account for the role of agents with international relationships in cluster adaptation. The role of internationalisation can be argued to lie with different search processes employed by local actors. In line with the argument proposed in Sect. 4.1) internationalisation would help agents find and evaluate configurations that are very different compared to their own with a greater accuracy. One way to account for this is to introduce an error term into the search process of agents without international linkages if they modify more than one or a small subset of their elements (see also

Levinthal 1997). Introducing a different amount of internationalised agents into the cluster then allows for further insight into the trade-off between local interactions and strong agglomeration externalities on the one as well as international openness on the other hand (Bathelt et al. 2002). In a sense, this kind of modelling exercise would introduce agent heterogeneity (with respect to their search processes) into the framework, thereby addressing the question of *optimum heterogeneity* in a very specific context.

A more long-term project regards the inclusion of more evolutionary aspects into the model. These could come from two possible sources. On the one hand, emergence and selection dynamics at the level of cluster agents would be required to allow for an analysis of the more long-term processes following a change event. Within this long-term perspective, the architecture of the cluster would also cease to be fixed, i.e. agent activities and strategies might change. Changes in agent activities would then alter the extent of division of labour in the cluster while modifications in agent orientation latter affect the mode of co-ordination (Galaskiewicz and Wasserman 1981). It will remain to be seen if cluster agents can also self-organise their activities in order to find optimal, i.e. very adaptive architectures. Such an inclusion of evolving architectures would however imply moving from the original N/K conceptualisation to the idea of evolving fitness landscapes for the system (Altenberg 1995; see Sect. 6.1.1).[125]

9.5 Conclusion

"It is not the strongest of the species that survive, nor the most intelligent, but the one most responsive to change" (Darwin 1859). These words very well sum up the overall purpose of this study, which set out to address the question of whether, how and when agents in existing clusters can successfully adapt to and thereby survive change events in their environment. It was found that differences regarding the extent of division of labour and the way agents co-ordinated their activities were important for cluster responsiveness to change. As a consequence, cluster adaptation and survival is a path dependent process where architectural features having evolved throughout the cluster's past history may make it more or less likely to survive future challenges. Regarding the question posed in the title ("A life cycle for clusters?") the study has shown that there is no determinism in the dynamics of agglomeration, change and adaptation. In some instances, decline may be unavoidable. However, surviving clusters also show that adaptation is a possibility. Among the different possible stages and levels of analysis involved with cluster survival, this study has focussed on their architectural properties that help or hamper the self-organisation process between its agents geared at adjusting to the new situation. By focussing on the short-term adaptation to change, the study thus shows when clusters are more likely to start off on a positive develop-

---

[125] An application of the Altenberg model to evolutionary questions in technological development can be found in Frenken 2005 as well as Frenken and Nuvolari 2004.

ment trajectory after the change event. Nonetheless, a good start does not guarantee survival for the cluster.

Against the background of severe structural problems in old industrial areas and the increasing pace of technological, market and political change characterising the modern economy, the research question of this study, i.e. the issue of how, whether and when agents in clusters can adapt to change grows in importance. Bearing in mind that the benefits to co-location advanced in the cluster concept occupy an increasingly central position in the interest of public policy as well as empirical and certainly also theoretical research, one has to investigate – given the existing evidence on the phenomenon of decline – when the advantages to being in an area can turn into liabilities as well as what can be done to help agents adapt to external challenges.

While the insights generated by the model are partial and subject to methodological constraints, they can still assist empirical work as well as political initiatives. Regarding empirical work, the study has set out to derive general causalities between division of labour, co-ordination and cluster adaptability. While a direct test for both aspects may only be possible under specific conditions, the findings do point at useful factors to be investigated. Put differently, empirical researchers could use the model results as a roadmap when starting out their analysis, which might at times be more helpful than starting a (case) study on clusters, change and adaptation without any idea of the 'terrain' upon which the success factors for cluster survival map out. Policy initiatives in turn might benefit from the model's results in two ways. First, the findings presented here can help actors identify whether certain structures and developments in a specific cluster are more likely to help its adaptability or not. Second, the model advances some caution with respect to an implicit goal underlying many cluster initiatives. In the context of creating and strengthening clusters, there seems to be an assumption that greater interaction between agents will foster their interdependence and thus their spatial embeddedness. With respect to agglomeration externalities, there thus seems to be a 'more is better' perspective. While very interdependent clusters can be highly competitive in specific environmental conditions (e.g. because they located one of the few high optima in the fitness landscape), interdependence hampers adaptability. As a result, generating as many externalities through policy initiatives (if this can be achieved at all) may be an unsustainable strategy in the long-run.

By showing that thanks to advances in complexity science, a Marshallian understanding of clusters is no longer at odds with a dynamic perspective, this study has contributed to the theoretic literature by taking a first important step into the direction of an evolutionary theory of clusters, change and adaptation (Boschma and Frenken 2005; Boschma and Lambooy 1999). It would be hoped that future research will look in more detail into the theoretic foundations that could underlie the existing variety of empirical evidence on clusters, change and adaptation in order to derive general causalities and thereby contribute to a better understanding of why some clusters survive while others do not. Moreover, the complex system perspective adopted here sheds hope that once one begins to understand the microdynamics, this may open up an avenue towards addressing foundations of the macrobehaviour in larger scale spatial constructs (Schelling 1978), where clusters are no the system but one of its elements. Understanding cluster dynamics might then be a start into investigating the development of their host entities such as cities, regions and maybe even nation states. Summing up, it is hoped that the study will assist theoretic effort in reviving the dynamic perspective on clusters change and adaptation that was prominent in the early days of location choice theory.

Appendices

A1 Symbiosis, habitat and change

The basic dynamics of symbiosis between populations in biology argue that the growth of a population depends on its own reproduction, symbiosis with the other population and habitat constraints (Murray 1993, pp. 83-84).

$$\frac{dN_1}{dt} = r_1 N_1 + \alpha_1 N_1 N_2 - l_1 N_1^2 \qquad \text{(A1.1)}$$

$$\frac{dN_2}{dt} = r_2 N_2 + \alpha_2 N_1 N_2 - l_2 N_2^2 \qquad \text{(A1.2)}$$

$r_x N_x$ Growth rate of population one and two (reproduction)
$\alpha_x N_1 N_2$ Symbiosis between populations one and two
$l_x N_x^2$ Growth constraint due to habitat conditions

With the adjustments described in Sect. 2.1, the development of the cluster's firm population (Eq. A1.3) is conditional on its own growth (reproduction) and the symbiotic effects between the population and itself (f), local conditions (c) and a supporting industry (s) all representing the effects of agglomeration economies.

$$\frac{df(t)}{dt} = a_{ff} \cdot f(t)^{\alpha_{ff}} + a_{ef} \cdot (\hat{f}(e) - f(t)) + a_{cf} \cdot c(t)^{\alpha_{cf}} + a_{sf} \cdot s(t)^{\alpha_{sf}} - \phi_f \cdot f(t)^{\rho_f} \qquad \text{(A1.3)}$$

$$\frac{dc(t)}{dt} = a_{fc} \cdot f(t)^{\alpha_{fc}} - \phi_c \cdot c(t)^{\rho_c} \qquad \text{(A1.4)}$$

$$\frac{ds(t)}{dt} = a_{fs} \cdot f(t)^{\alpha_{fs}} - \phi_s \cdot s(t)^{\rho_s} \qquad \text{(A1.5)}$$

The term $a_{ff} \cdot f(t)^{\alpha_{ff}}$ and its' equivalents in Eqs. A1.3 to A1.5 describe the symbiotic relationship between the different variables and the growth of the local firm population, while the habitat constraints are expressed as terms of the form $\phi_f \cdot f(t)^{\rho_f}$. In addition, $a_{ef} \cdot (\hat{f}(e) - f(t))$ accounts for the effect of external conditions on the local firm population by reflecting that its state is driven towards the area's carrying capacity. Parameters α_{ff} α_{cf} α_{sf} α_{fc} α_{fs} denounce the minimum levels of c, f and s for the emergence of symbiotic processes (agglomeration economies) whereas ρ_f ρ_c ρ_s indicate the maximum levels of the respective variables before growth constraints within the region begin to set in. If the symbiotic interactions are to be relevant, they need to become effective before growth constraints interfere, i.e. it is assumed that α_{ff}, α_{cf}, α_{sf}, α_{fc}, α_{fs} < ρ_f, ρ_c, ρ_s.

Mathematically, the resulting system of equations yields two different states depending on exogenous conditions (e) and parameter values. The first state is characterised by one stable equilibrium value for the number of local firms. If there are n regions hosting clusters of the industry described in this fashion, they all converge towards this equilibrium, implying a uniform industry distribution. The second state knows two equilibria with low or high numbers of firms in each area and thus allows for the emergence of an agglomeration if some regions converge to the first and some to the second equilibrium. Mathematically, the second state only arises if one of the parameters for the symbiotic relationships has a value larger than one, i.e. the respective relationship is relatively weak if there are few local firms but becomes over proportionally strong as the size of the cluster increases. One example of such a relationship is policy support that will likely be stronger for large than for small local industries. It is in the context of this second state, that the development of clusters can be analysed.

A2 Clusters as core-periphery structures

A key aspect underlying the models of the New Economic Geography regards the consumer's utility function $U = C_M^{\mu} \cdot C_A^{1-\mu}$ where 1-μ is the share of manufacturing in consumer expenditure. In it, the utility from consumption of manufacturing products relates to the number of varieties c_i and the quantities of each variety consumed:

$$C_M = \left[\sum_{i=1}^{N} c_i^{(\sigma-1)/\sigma} \right]^{\sigma/(\sigma-1)} \qquad (A2.1)$$

$\sigma > 1$ Elasticity of substitution between varieties

Iceberg transport costs are included by expressing that of each unit shipped between regions only a fraction $\tau < 1$ arrives. The larger τ, the lower transport costs. Producers maximise profits by setting the price for the local and non-local products in region one or two as:

$$p_{1/2} = \left(\frac{\sigma}{\sigma - 1} \right) \cdot \beta w_{1/2} \qquad (A2.2)$$

$$P_{1(2)/2(1)} = \frac{P_{1/2}}{\tau} \qquad \text{(A2.3)}$$

$w_{1/2}$ Wages in region one and two

Due to free entry in the sector, profits are driven to zero. This means that output per firm in each region is identical, regardless of wage rates, relative demand and the likes. Therefore, industry output in each region hinges only on the latter's endowment with labour and the distribution of the industry between both regions is conditional on the distribution of manufacturing labour $L=L_1+L_2$ over space. A given distribution of labour (without factor mobility) then yields the short-term equilibrium values for wages, income and so on. Workers are then argued to move to the region offering them higher real wages, which leads to an agglomeration or dispersion of the industry.

To investigate the role of different model parameters, Krugman assumes that all workers (and firms) are concentrated in region one. The question is then, when it pays for one defecting firm to start producing in the other region, thus enabling a dispersion of the industry over space. Transportation costs work to the firm's disadvantage when selling to region one but constitute an advantage for sales in region two. In addition, wages for manufacturing workers are higher in region two as workers have to be attracted there by higher salaries. The breaking point of where relocation becomes profitable is defined as the quotient of the value of sales in region two and that of sales in region one.

$$\frac{V_2}{V_1} = v = \frac{1}{2}\tau^{\mu\sigma}\left[(1+\mu)\tau^{\sigma-1} + (1-\mu)\tau^{-(\sigma-1)}\right] \qquad \text{(A2.4)}$$

Put differently, if $v>1$, relocation becomes profitable and the fully concentrated equilibrium is unstable. The value of v is then related to different parameter values for manufacturing expenditure, transport costs and scale economies in order to determine the "breaking points" of the fully concentrated equilibrium.

$$\frac{\partial v}{\partial \mu} = v\sigma(\ln \tau) + \frac{1}{2}\tau^{\sigma\mu}\left[\tau^{\sigma-1} - \tau^{-(\sigma-1)}\right] < 0 \qquad \text{(A2.5)}$$

The larger the share of income spent on manufacturing products, the stronger the attractiveness of the large market in region one and workers demand a larger wage premium to move to the other area. In consequence, sales of a defective firm are lower. "*A larger share of manufactures in consumer expenditures also favours agglomeration, because it augments the impact of immigration on the size of the local market for manufactures. In addition, it increases the weight of the prices of manufactures in real wages, thus enabling firms located in regions with more industry to attract workers without having to pay high nominal wages*" (Ottaviano and Puga 1997, p. 10).

$$\frac{\partial v}{\partial \tau} = \frac{\mu \sigma v}{\tau} + \frac{\tau^{\mu\sigma}(\sigma-1)\left[(1+\mu)\tau^{\sigma-1} - (1-\mu)\tau^{-(\sigma-1)}\right]}{2\tau} \qquad (A2.6)$$

For high transport costs (low levels of τ), defection is profitable. As transport costs decrease, the value of defection becomes smaller than 1 for intermediate transport costs. Further decreases in transport cost then see the value of defection approach 1 from below as location becomes increasingly irrelevant.

$$\frac{\partial v}{\partial \sigma} = \ln(\tau)\left\{\mu v + \frac{1}{2}\tau^{\mu\sigma}\left[(1+\mu)\tau^{\sigma-1} - (1-\mu)\tau^{-(1-\sigma)}\right]\right\} = \ln(\tau)\left(\frac{\tau}{\sigma}\right)\left(\frac{\partial v}{\partial \tau}\right) \qquad (A2.7)$$

The higher scale economies (the lower elasticity of substitution; Krugman 1991b, p. 490), the lower the profits from defection since it pays very strongly to concentrate production in one site and the attractiveness of region one bears out very strongly in this context (sales of local products are higher due to their lower cost implying that firms near a large market make greater use of these scale economies). In addition, "*[...] a lower elasticity of substitution across varieties in consumers' preferences increases the importance of having a large variety of products available locally. By reinforcing the monopoly power of firms over their own varieties, this weakens local competition and favours agglomeration*" (Ottaviano and Puga 1997, p. 10).

A3 Location choice, path dependence and clusters

The location benefits of an area (B_{fq}) are argued to depend on an area's intrinsic features and firm's preferences (G_{fq}) on the one, as well as agglomeration benefits (A_{fq}) on the other hand.[126]

$$B_{fq} = G_{fq} + A_{fq}(n_q) \qquad (A3.1)$$

Depending on the benefits obtained (B_{fq}), each firm has a certain preference for each location. At a given point in time, it will locate in region q if its benefits in q (geographic and agglomeration returns) exceed those of all other areas (i), i.e. $B_{fq} > B_{fi}$. The distribution of firms' locational preferences (influencing G_{fq}) is random in order to introduce 'historical accident'. Therefore, the probability that a firm preferring location q over all others entering in the next time period yields:

---

[126] If $A_{fq}(n_q)$ is increasing in n_q, agglomeration economies exists, if it decreases, there are diseconomies from agglomeration (also termed 'congestion cost').The notation employed here differs from that used by Arthur himself. It has been adapted to that of the model introduced by Maggioni 2002 to ensure better comparability.

$$p_q = \text{Pr } ob \left\{ \left\lfloor G_{fq} + A_{fq}(n_q) \right\rfloor > \left\lfloor G_{fi} + A_{fi}(n_i) \right\rfloor all \rightarrow i \neq q \right\} \qquad \text{(A3.2)}$$

This probability also depends on the distribution of firms having entered the regions before as they determine each area's agglomeration benefits ($A_{fx}(n_x)$).

The number of firms in each location ($n_q = n_q/n$) or the share of each location of the (total number of firms in the) industry (x_q) then mirrors its spatial distribution.

$$n_q = (n_1(n),\dots, n_Q(n)) \text{ and } x_q = (x_1(n),\dots, x_Q(n)) \qquad \text{(A3.3)}$$

Since the industry's spatial distribution evolves with one firm being added at a time, the number of firms in any location at a point in time equals the number of firms already located there plus the chance of a firm with a preference for this area being the next to chose its site.

$$q_{n+1} = n_q + b(n; x_q) \qquad \text{(A3.4)}$$

Where b in Eq. A3.4 is a unit vector assigning the next entering firm to one of the possible locations. The probability of the next firm choosing location q (i.e. b being 1 in location q) equals pq(n;xq) and depends on q's share of the industry (xq) as well as the total number of firms in the sector (n).

Accordingly, the share of each region in the total number of firms evolves as:

$$x_{n+1} = x_q + \frac{1}{n+1}\left[b(n; x_q) - x_q\right], \text{ or} \qquad \text{(A.3.5)}$$

$$x_{n+1} = x_q + \frac{1}{n+1}\left[p(n; x_q) - x_q\right] + \frac{1}{n+1}\mu(n; x_q), \text{ with}$$
$$\mu(n; x_q) = b(n; x_q) - p(n; x_q)$$

The locational shares of all areas are therefore dependent on the attractiveness of the region, which drives the expected motion of the firms in the industry. At the same time, such a stabilisation is perturbed by the randomness of firm entry ($\mu(n;x_q)$). As both motion and perturbation effects become smaller with growing firm numbers in the industry, locational shares can be expected to settle at a given level if the industry becomes large. Uhus nder certain conditions,[127] there are different combinations of industry allocations over all areas, which are stable, fixed points. These fixed points are shaped (regarding the industry share of each area) by the locational probability function (p); i.e. they depend on the differentials in

---

[127] More detail on the mathematical foundations is provided in Arthur 1990, p. 240. It is argued that that their notion of equilibrium shares in a path dependent process is a generalisation of the strong law of large numbers where shares are stabilised if increments are added independently of the present state of the process.

firms' locational tastes as well as the impact of agglomeration effects on an area's profitability.

Gross locational benefits B_{fq} for firm f locating in area q are again a composite of geographic and agglomeration benefits.

$$B_{fq} = G_{fq}(k_q, l_q, s_q, u_q) + A_{fq}(n_q) \qquad (A3.6)$$

- G_{fq} = geographical benefits: Depend on intrinsic features of the area, e.g. the quality of local production factors (capital k_q and labour l_q), the efficiency of the local network of specialised (service) suppliers s_q, urban and industrial infrastructure u_q.
- $A_{fq}(n_q)$ = agglomeration benefits: A concave, non-monotonic function of n_q implying that gross benefits initially increase with the number of firms due to agglomeration economies but decrease once an optimum cluster size in terms of firm numbers has been reached.

Locational cost c_{fq} are can correspondingly be expressed as the sum of geographic and agglomeration cost.

$$c_{fq} = g_{fq}(r_q, w_q, d_q, t_q) + a_{fq}(n_q) \qquad (A3.7)$$

- g_{fq} = geographical cost: Reflect the cost structure of the area, i.e. local interest rates r_q, wages w_q, average service prices d_q, land rents and taxation t_q.
- $a_{fq}(n_q)$ = agglomeration cost/ congestion cost: A convex, non monotonic function of n_q, i.e. local costs initially decrease with firm numbers due to agglomeration economies but start to increase after an optimum level as the greater competition for a limited pool of inputs and infrastructure increases their price.

As a consequence, net locational benefits Net_{fq} can be expressed as the differences between total benefits and costs. As long as Net_{fq} is positive, firms will locate in the area, increasing the size of the agglomeration.

$$Net_{fq} = B_{fq} - c_{fq} = H_{fq}(r_q, w_q, d_q, t_q, k_q, l_q, s_q, u_q) + h_{fq}(n_q) = a_q - h_{fq}(n_q)^{128} \quad (A3.8)$$

Since Net_{fq} constitutes the difference between a concave and a convex function, it is concave itself, implying that up to a threshold level, each new firm entering the agglomeration increases the average profitability of locating in the region. Beyond the threshold level, new entrants decrease the average net benefits available to both incumbents and new entrants.

---

[128] If geographic benefits and costs do not change over time, H_{fq} can be summarised by a constant term α_q.

The development process is modelled stressing the relevance of firms' spatial interactions, i.e. *"the rate of growth of the industrial mass [of the cluster] equals the product of the individual firm's contribution to the regional population's growth [i.e. its contribution to local benefits and costs] and the number of firms already in the region"* (Maggioni 2002, pp. 100f). This is expressed by a logistic equation.

$$\frac{dn_q}{dt} = r_q\, n_q(t) \left(1 - \frac{n_q(t)}{K_q} \right) \qquad\qquad (A3.9)$$

The firm's contribution to regional growth (net locational benefits) decreases as a linear function of regional population size. This leads to the aforementioned S-shaped curve: as the area is small ($n_q(t)$ is near zero), the growth constraint of a firm's entry $\left(\frac{n_q(t)}{K_q} \right)$ is close to one, implying that the area develops as if only agglomeration economies existed (at a rate of $r_q n_q(t)$). As the number of firms approaches the maximum level K_q, the constraint on growth becomes more apparent, i.e. $\left(\frac{n_q(t)}{K_q} \right)$ approaches zero.

A4 Model parameter values and fitness landscapes

EvenK=4   K= 2.33   C= 0.67 *

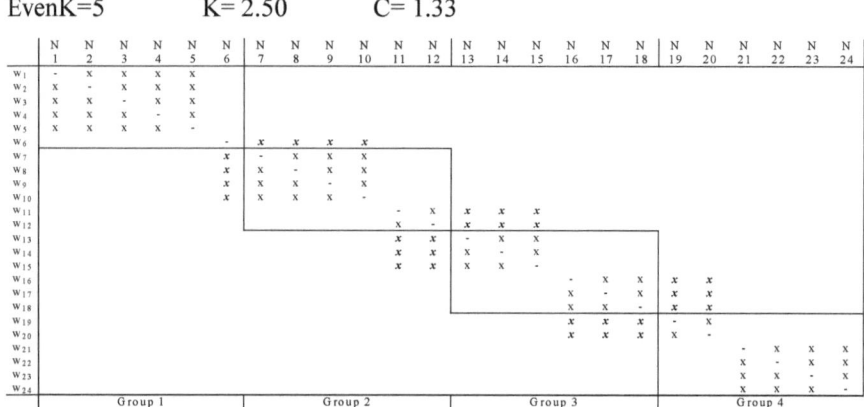

	N1	N2	N3	N4	N5	N6	N7	N8	N9	N10	N11	N12	N13	N14	N15	N16	N17	N18	N19	N20	N21	N22	N23	N24
W1	-	x	x	x																				
W2	x	-	x	x																				
W3	x	x	-	x																				
W4	x	x	x	-																				
W5					-	x	*x*	*x*																
W6					x	-	*x*	*x*																
W7					*x*	*x*	-	x																
W8					*x*	*x*	x	-																
W9									-	x	x	x												
W10									x	-	x	x												
W11									x	x	-	x												
W12									x	x	x	-												
W13													-	x	x	x								
W14													x	-	x	x								
W15													x	x	-	x								
W16													x	x	x	-								
W17																	-	x	*x*	*x*				
W18																	x	-	*x*	*x*				
W19																	*x*	*x*	-	x				
W20																	*x*	*x*	x	-				
W21																					-	x	x	x
W22																					x	-	x	x
W23																					x	x	-	x
W24																					x	x	x	-

Group 1 | Group 2 | Group 3 | Group 4

* K and C values are calculated taking the total number of intra- (denoted by x) and inter-agent (denoted by *x*) externalities divided by the number of elements (N=24). In the case of EvenK=4, this yields 56 intra-agent externalities (i.e. K=56/24=2.33) and 16 inter-agent externalities (i.e. C=16/25=0.67). The following K and C values are determined accordingly.

EvenK=5   K= 2.50   C= 1.33

	N1	N2	N3	N4	N5	N6	N7	N8	N9	N10	N11	N12	N13	N14	N15	N16	N17	N18	N19	N20	N21	N22	N23	N24
W1	-	x	x	x	x																			
W2	x	-	x	x	x																			
W3	x	x	-	x	x																			
W4	x	x	x	-	x																			
W5	x	x	x	x	-																			
W6						-	*x*	*x*	*x*	*x*														
W7						*x*	-	x	x	x														
W8						*x*	x	-	x	x														
W9						*x*	x	x	-	x														
W10						*x*	x	x	x	-														
W11											-	x	*x*	*x*	*x*									
W12											x	-	*x*	*x*	*x*									
W13											*x*	*x*	-	x	x									
W14											*x*	*x*	x	-	x									
W15											*x*	*x*	x	x	-									
W16																-	x	x	*x*	*x*				
W17																x	-	x	*x*	*x*				
W18																x	x	-	*x*	*x*				
W19																*x*	*x*	*x*	-	x				
W20																*x*	*x*	*x*	x	-				
W21																					-	x	x	x
W22																					x	-	x	x
W23																					x	x	-	x
W24																					x	x	x	-

Group 1 | Group 2 | Group 3 | Group 4

EvenK=7   K= 3.17   C= 2.33

EvenK=8   K= 3.67   C= 3.33

EvenK=9   K= 4.25   C= 3.00

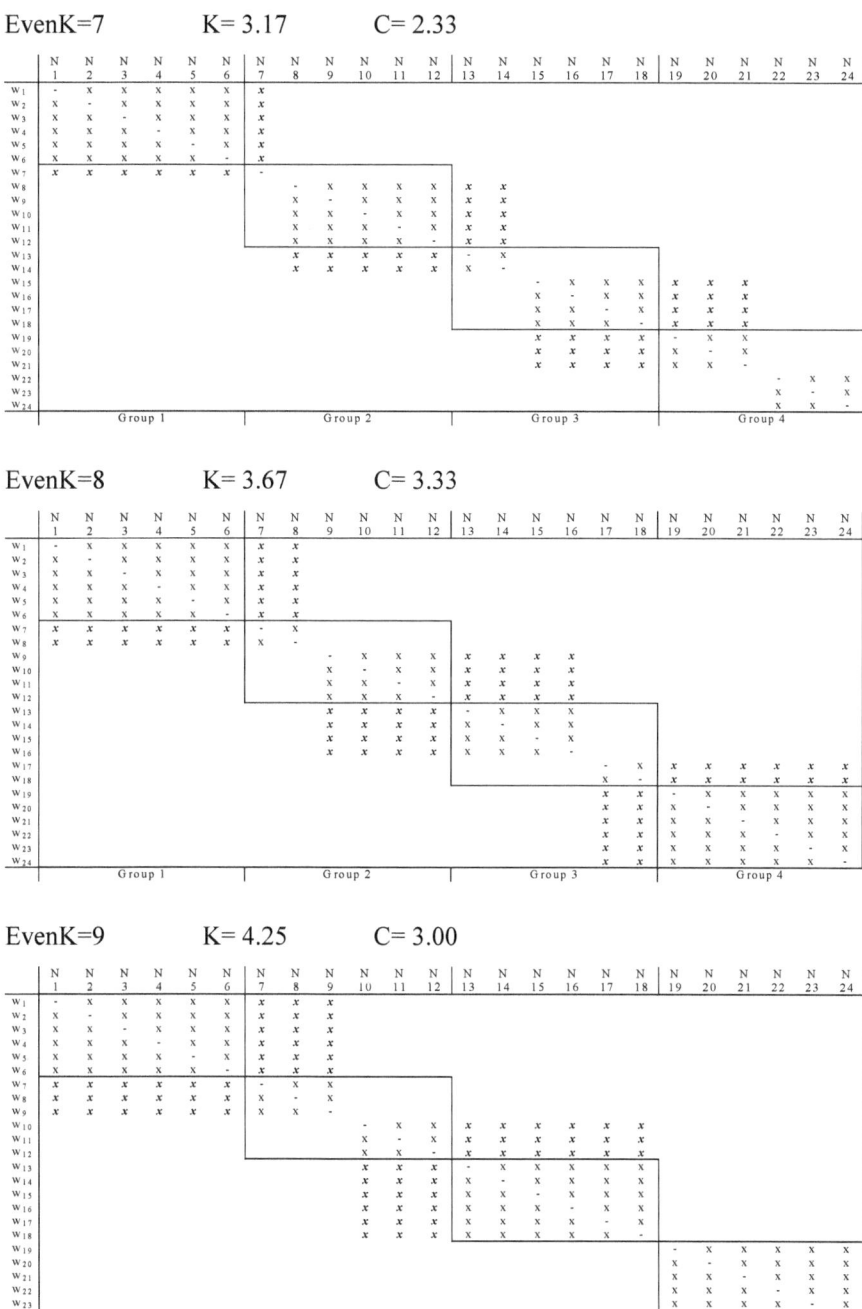

EvenK=10   K= 3.67   C= 4.33

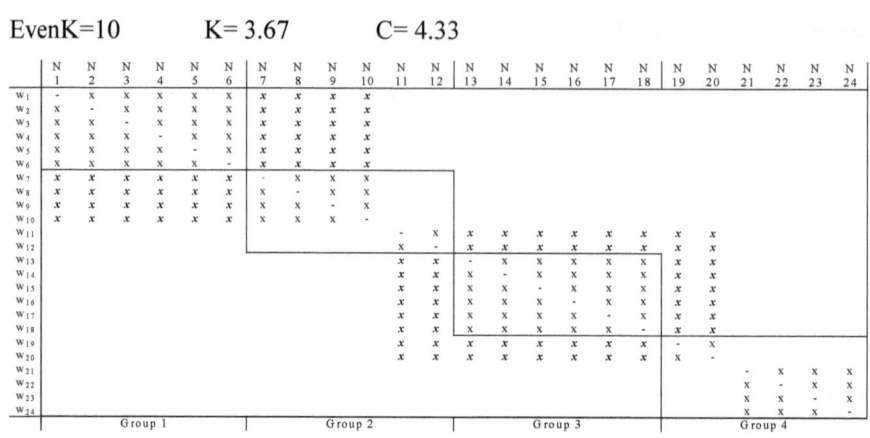

EvenK=11   K= 3.92   C= 5.33

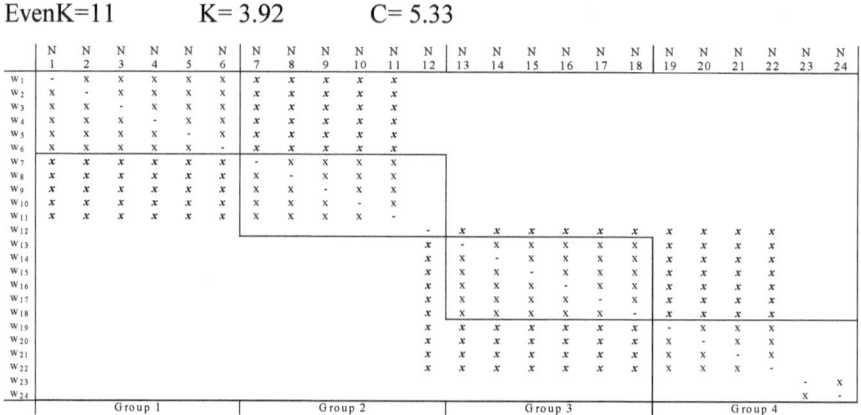

EvenK=12   K= 5.00   C= 6.00

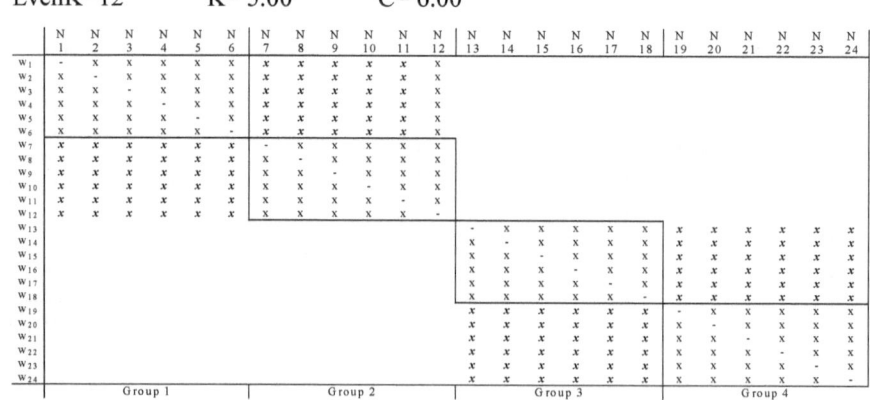

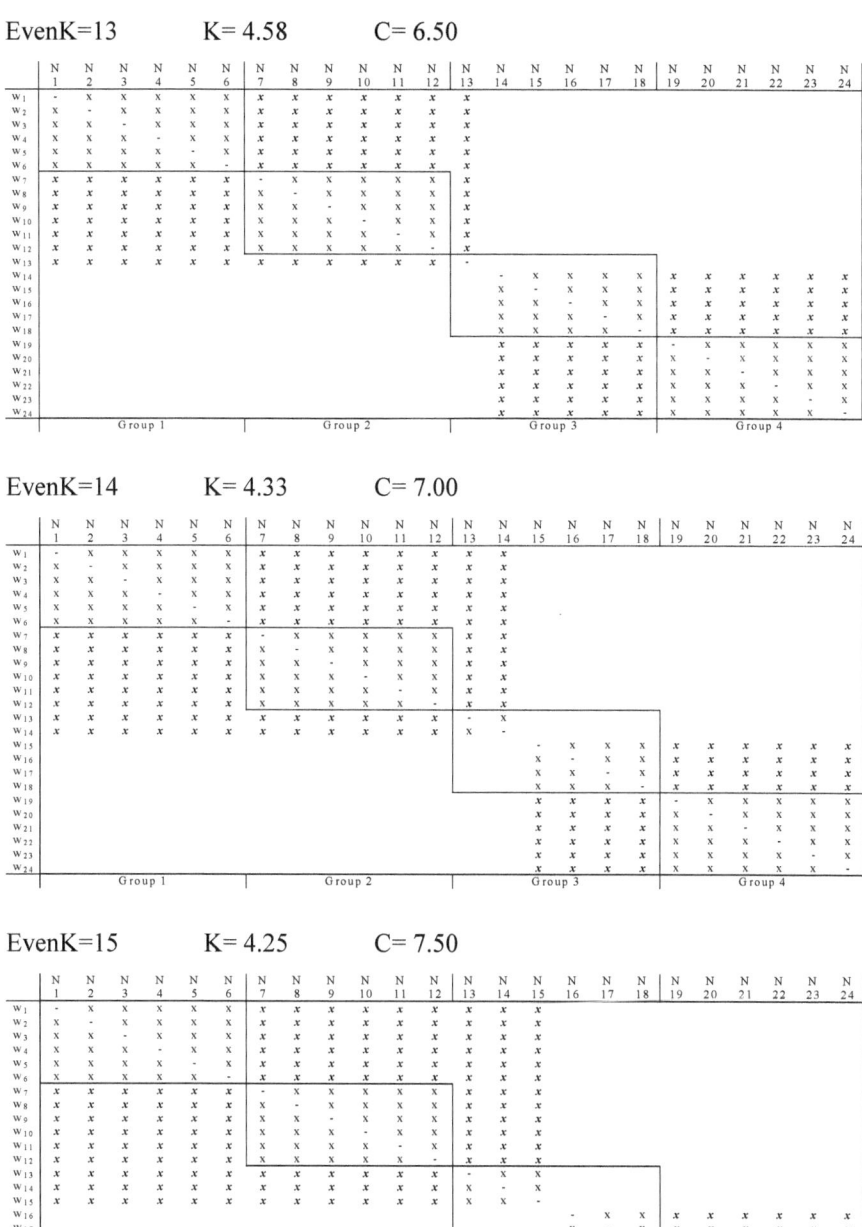

EvenK=16   K= 4.33   C= 8.00

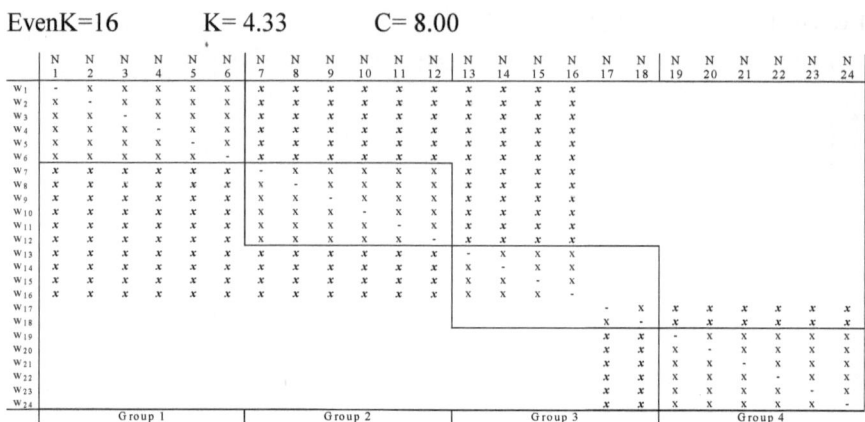

EvenK=17   K= 4.58   C= 8.50

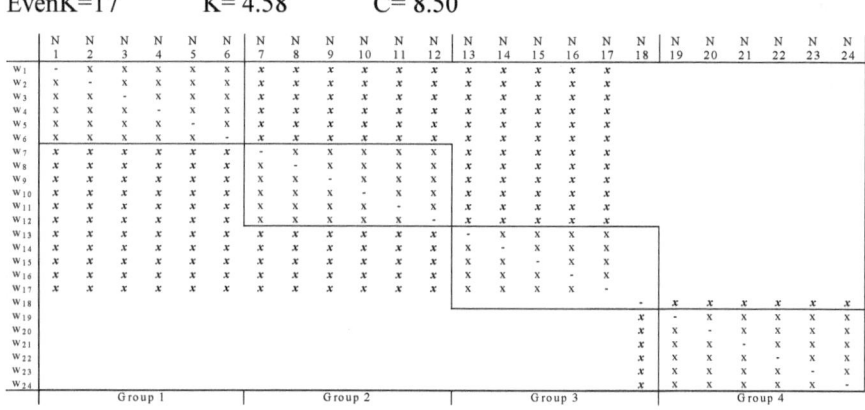

EvenK=18   K= 5.00   C= 9.00

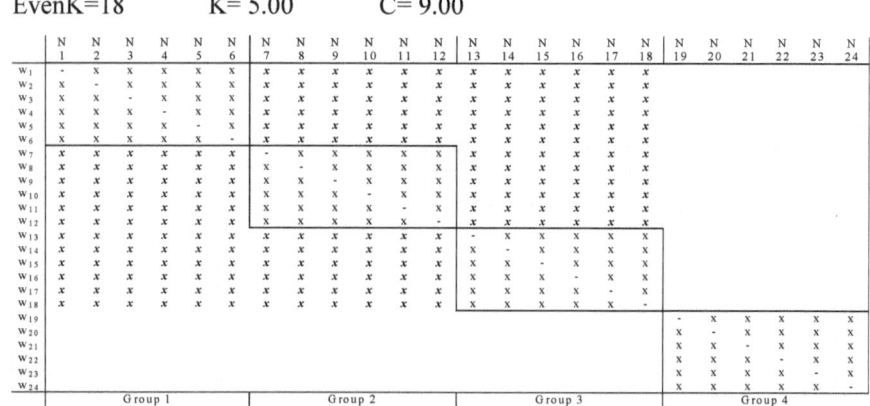

EvenK=19   K= 4.58   C= 10.5

EvenK=20   K= 4.33   C= 12.00

EvenK=21   K= 4.25   C= 13.50

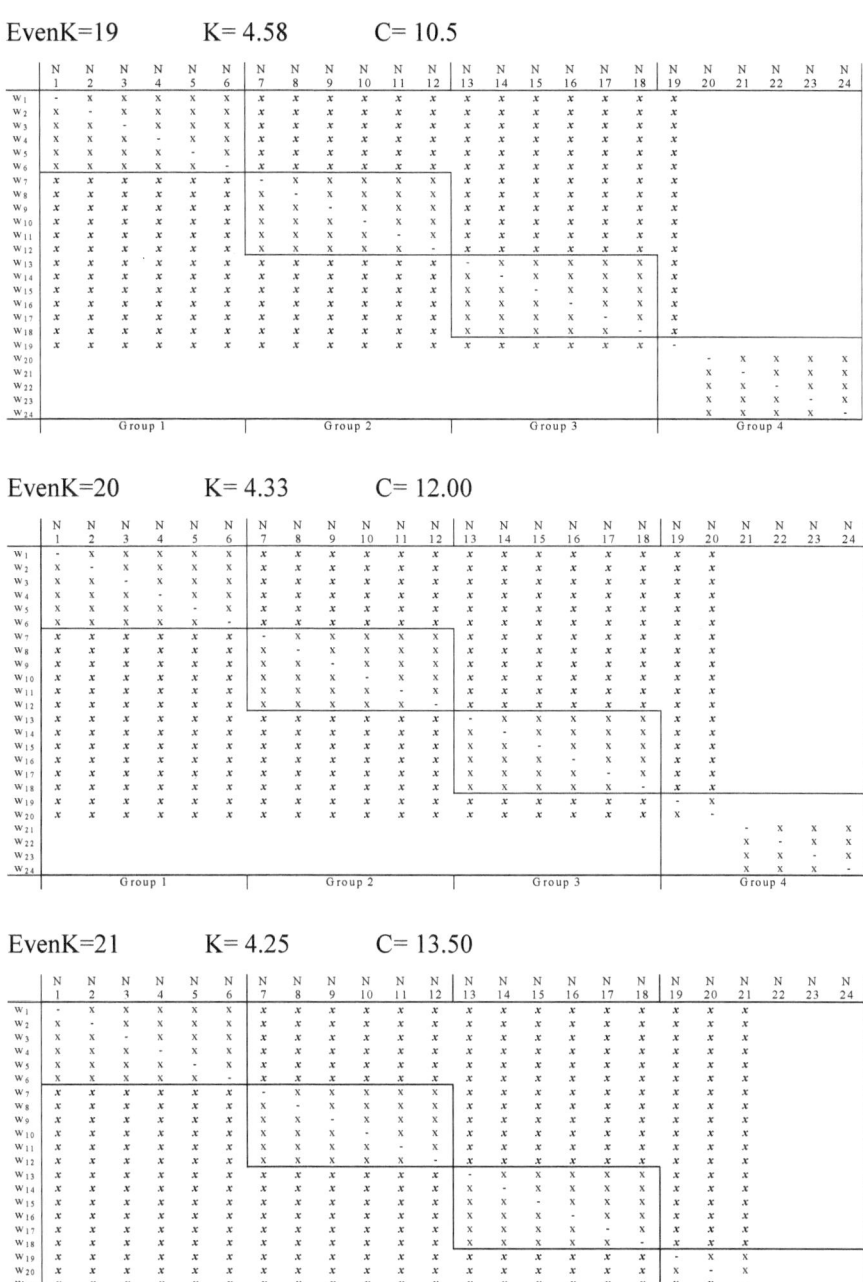

EvenK=22   K= 4.33   C= 15.00

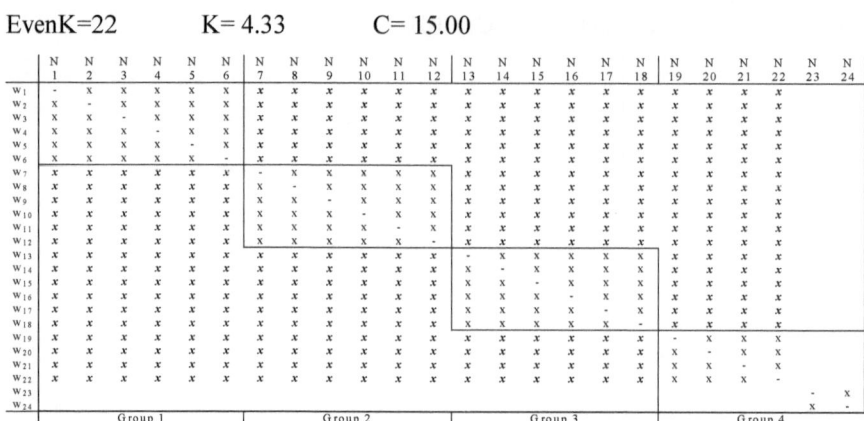

EvenK=23   K= 4.58   C= 16.50

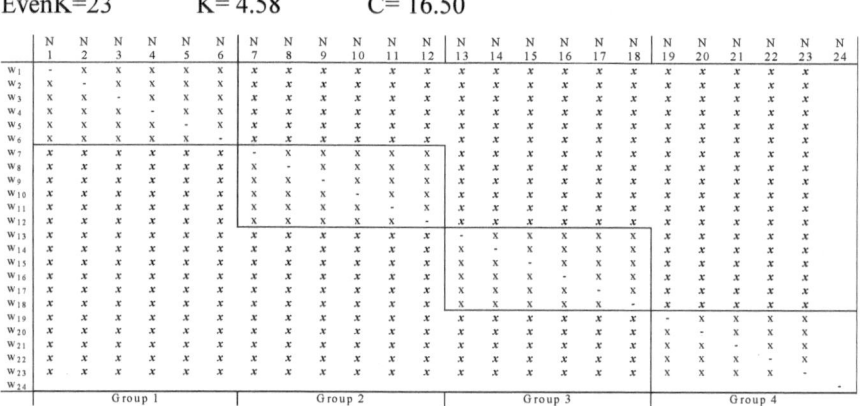

EvenK=24   K= 5.00   C= 18.00

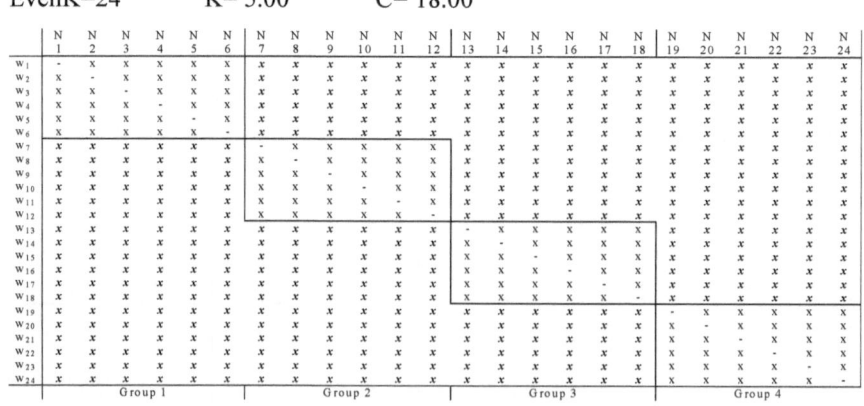

A5 The simulation model

As is depicted in table A5.1., the simulation model presented here consists of two separate units. One (FF) represents the entire fitness landscape whereas the other (Population) mirrors the different clusters. Put differently, the object "Population" accounts for the differently architectured clusters (with respect to their mode of co-ordination that is reflected in agents' selection mechanisms, see also table A5.2.) while the object "FF" provides a measure of success of their adaptive measures within each simulation run. Changing the relevant parameters reflecting element interdependence in the object "FF" (*EvenK, ForeOverlap, AftOverlap*) furthermore allows for an indirect inclusion of the effects of a growing or decreasing division of labour on cluster performance (For additional detail on objects, *parameters* and variables, as well as simulation settings see tables A5.3.-A5.5.).

Table A5.1. Model object structure

Root		
FF "FF" contains all objects, variables and parameters making up the fitness landscape	Population "Population" contains all objects, variables and parameters making up the cluster and its agents. Populations differ according to their agents' selection mechanisms representing different forms of co-ordination in the cluster.	
Bit "Bit" represents all the objects, variables and parameters working upon one of the N=24 elements in the fitness landscape.	Substring "Substring" denotes the number of decompositions of the entire string controlled by the agent groups in the cluster. Here Substring=4.	
Link "Link" represents all interdependencies between system elements N. Links can exist between elements within the control of one agent (K) as well as between agents (C).	SSBit "SSBit" identifies the n elements pertaining to a substring. SSBit=6	Agent "Agent" represents all actors conducting activities in the cluster. Agents are allocated into groups pertaining to a given substring.
		ABit "ABit" denotes all the activities n under the control of one agent. Here ABit=6.

At the start of each simulation run, the fitness landscape (FF) is initialised using the equation Init. Starting from their (randomly assigned) initial configurations, Agents are then allowed to explore their subset of the fitness landscape through mutation and selection. More precisely, agents change all six ABits under their control with a given probability (*ProbMut=0.5*) using the variable Mutation. They then evaluate the expected local fitness (*ExpLocFit*) given by the variable TestMe of their new configuration (*TValues*) in comparison to the previous one (*AValues*). The extent to which elements in the entire string enter the agents' decisions by influencing the TestMe function and are expressed in the expected local fitness

(*FlagTest*) reflects the different modes of co-ordination in the cluster. In individualistic clusters, only the elements under each agent's control enter the evaluation of her strategy. In the collective case, all elements enter an agent's selection mechanism whereas in alliance scenarios, only the agent's and the alliance partner's fitness matter. In the dominant firm scenario, all agents aim at improving their own and the dominant actor's fitness (except for the latter who behaves egoistically). The expected local fitness is derived by holding the states of the elements outside an agent's control constant.

To get from individual agent to agent group dynamics (SubString), a bidding process is introduced where agents bid with their new configuration in Bidding. Agents are chosen to represent their group (their substring) based on a certain *Criterion*. Here, the *Criterion* applied for an agent to be *Chosen* as the representative of her substring is the highest *ExpGlobFit* derived by TestMe (again holding the rest of the string constant). If an agent is chosen to represent his substring, the current configuration of her bits (*ABit*) becomes the configuration of the substring bits (*SSBit*). Agent groups are then aggregated into cluster dynamics (Population) by taking the configurations of all substrings and computing the new fitness values for agent groups (*ActLF*) and the cluster (*Fitness*) through FitFun. Both values are saved for each simulation step. Having derived the actual fitness values for both, the process starts again with individual agents attempting a mutation of their current configuration.

Depending on whether a bifurcation event or an environmental perturbation is investigated, simulations are changed every 600 steps. In the case of bifurcations, the Populations are reinitialised with new fitness landscapes (FF) using the function Init. In the case of perturbations, the fitness landscape is changed for *NumShiftBit* elements every 600 steps using the Shift function.

Table A5.2. Agent orientations (reflecting different modes of co-ordination)

	Selection mechanism (FlagTest)
Individualistic	Each agent cares only for her own fitness, i.e. for the fitness of the elements under her control (FlagTest=FlagSString=6).
Collective	Each agent cares for the fitness of the entire cluster, i.e. all variables enter her evaluation of the strategy (FlagTest=N=24).
Alliance	Each agent cares for her own and ally fitness, i.e. FlagTest=12. The distribution of these elements then depends on the constellation of alliances in the cluster.[129]
Leader	The dominant firm cares only for its own fitness (FlagTest=FlagSString=6). The other agents in the cluster care for their own fitness and that of the dominant firm, i.e. their FlagTest=12.

Agent orientations matter as they determine what elements enter the TestMe function (FlagTest=1 or 0) evaluating a given strategy by returning the agents expected local fitness.

Table A5.3. Simulation settings

Label	Features
SString6, Agent 2(1)	Nine populations representing individualistic, collective as well as different alliance and hierarchy clusters containing two agents each, except for the case of the dominant firm where the agent group consists of one agent only. Used to test the effect of different numbers of agents on cluster performance.
SString6, Agent 5(1)	Nine populations representing individualistic, collective as well as different alliance and hierarchy clusters containing five agents each, except for the case of the dominant firm where the agent group consists of one agent only. Used to generate results regarding the relative performance of different cluster co-ordination mechanisms (see table A5.2.) for changing degrees of interdependence.
SString6, Agent 5(5)	Nine populations representing individualistic, collective as well as different alliance and hierarchy clusters containing five agents each (the dominant firm is replaced by a dominant group containing five agents). Used to test the effect of a dominant agent group.
SString6, Agent 10(1)	Nine populations representing individualistic, collective as well as different alliance and hierarchy clusters containing ten agents each, except for the case of the dominant firm where the agent group consists of one agent only. Used to test the effect of different numbers of agents on cluster performance.
SString6, Indiv_Coll	Ten populations representing different constellations of egoistic and collectively oriented agent groups. Populations 1-4 exhibit one egoistic group in each of the four substrings, populations 5-10 have different constellations with two egoistic and two collective agent groups. Used to test the stability of the collective regime against egoistic behaviour.
SString6, Inv_Ego	Four populations representing different numbers of egoistic agents (1-4) in the first agent group. Used to test when the "prisoner's dilemma" begins to materialise.

---

[129] See also table A6.1. in App. 6.

Table A5.4. Objects, parameters and variables

Root

Label	Comment
Init (0)	Equation setting the basic initialisations for the fitness landscape as well as the different populations. Computed only once, it transforms itself into a parameter and is not computed again.
CreateFitContrib (0)	Variable in the fitness landscape that creates the fitness values. Initialises the vectors used for computing and storing the landscape fitness values. Appears in equation for: Init
N (P)	Parameter measuring the number of landscape elements (N=24). Appears in equation for: FitFun, CreateFitContrib, Init, Shift, InitEvenFF, InitKauffFF
NumFF (P)	Parameter measuring the number of fitness functions used in the simulation. Here, NumFF=1. Appears in equation for: FitFun, Init, Shift
PeriodShift (P)	Parameter measuring the number of simulation steps to be executed before the next shift in the fitness landscape. In the perturbation model, PeriodShift=600 steps. Appears in equation for: Shift
Shift (1)	Equation altering the fitness contributions for a limited number of landscape elements (NumShiftBit). Appears in equation for: Shift
UseMem(P)	Parameter determining whether fitness landscape values are stored for system configurations or not. Here, UseMem=0, i.e. landscape values are only stored for current system configurations. If the same point is revisited later in the simulation, its fitness is determined again using the fitness contribution values. Allows to implement N/K fitness landscapes even if their total number of points is larger than the available memory. Appears in equation for: Init
NumShiftBit (P)	Parameter measuring the number of elements whose fitness contribution is altered in each shift. Here, NumShiftBit=6 (in the perturbation model) Appears in equation for: Shift

Object FF
Contained in Object: Root
Containing Objects: Bit
Appears in equation for: Init, Shift

IdFF (P)	Parameter identifying each fitness function (only one used here). Appears in equation for: FitFun
InitEvenFF	Initialises the object structure for a fitness function with evenly (block) distributed interdependencies Appears in equation for: InitFF
InitKauffFF (0)	Initialises the object structure for a fitness function with randomly distributed interdependencies Appears in equation for: InitFF

Table A5.4. (Cont.)

TypeFF (P)	Determines whether a fitness function with evenly or randomly distributed elements is used (here, InitEvenFF) Appears in equation for: InitFF
InitFF (0)	Initialises the fitness functions Appears in equation for: InitFF

FitFun (0)	Computes the fitness for each element state. This implementation makes use of dynamically allocated memory. It means that FitFun stores in memory the fitness of points already computed, including the fitness contribution of each bit of the string. If the point has never been computed before, then a new fitness value is randomly generated respecting the constraints defined in the landscape and stored in memory for future uses. The core idea of NK systems is that the fitness of a binary point is computed as the average of the fitness contributions (fc) of each element. The epistatic relations of NK systems define which bit influences other bits. Therefore, the system must store in memory $2^{(K+1)}$ fc's for each bit linked to other K bits. The current implementation, summarised below, allows for the creation of as complex landscapes as desired, even if the whole NK system would require and impossibly huge amount of memory. The system continues to allocate memory for new points until the operating system's memory limitations are reached. At that point the LSD programme crashes. Technically, the data on fitness values are stored in a memory structure defined as:
	struct bit
	{int id; //id of the bit (from 1 to N)
	struct dynlink l; //see below
	int *link; //vector of integers reporting the bit's linked to the bit
	int nlink; //number of bit's linked to the bit (i.e. length of *link)
	double fitcontr; //fitness contribution of the bit, computed by FitFun whenever requested};
	struct dynlink {dynlink *l0; dynlink *l1; double *f; };
	The dynlink is an element of the binary linked chain generated whenever the fitness of a point is computed. Starting from bit.l a unary linked chain is generated passing for l0 or l1 depending on the states of the related bits. Only the last dynlink contains an existing f field, for the fitness contributions.
	When FitFun is requested it stores the states of related bits for each bit in *link. Then, in l checks whether l0 exists or is NULL (if the related bit is 0) or checks l1, if the related bit is 1. This continues until either:
	a NULL l0 or l1 is encountered. Continue to generate a new linked chain until the end of related bits and generate a new fitness contribution stored in f.
	the related bits are finished. Return the f value as fitness contribution.
	At the end of each simulation running the "close_sim()" function cleans up all the memory.
	Appears in equation for: TestMe, Fitness, Shift

Table A5.4. (Cont.)

EvenK (P)	Parameter indicating the number of elements that are reciprocally linked, i.e. it determines the size of the blocks of interdependent elements Appears in equation for: InitEvenFF, InitKauffFF
ForeOverlap (P)	Parameter indicating how many of the EvenK interdependent elements are unilaterally linked with elements in the previous block (first block elements linked with those in the last block). Appears in equation for: InitEvenFF, InitKauffFF
AftOverlap (P)	Parameter indicating how many of the EvenK interdependent elements are unilaterally linked with elements in the next block (last block elements linked with those in the first block). Appears in equation for: InitEvenFF, InitKauffFF

Object Bit

Contained in Object: Root → FF

Containing Objects: Link

Appears in equation for: FitFun, Shift, InitEvenFF, InitKauffFF

IdBit (P)	Parameter identifying each of the N elements (Bits) of the fitness landscape. Appears in equation for: TestMe, Fitness, Shift, InitEvenFF, InitKauffFF
FitContr (P)	Parameter denoting the fitness contribution of each element state as a result of the state of that element (Bit) and those of interdependent ones. Appears in equation for: FitFun, TestMe, Fitness, Shift
appBit (P)	Parameter tagging the n=6 Bits affected by the shift function. Appears in equation for: Shift

Object Link

Contained in Object: Root → FF → Bit

Appears in equation for: FitFun, Shift, InitEvenFF, InitKauffFF

IdLink (P)	Parameter identifying which two elements are connected (existing link). Appears in equation for: InitEvenFF, InitKauffFF
Exist (P)	Parameter showing whether a link between two elements exists (1) or not (0). Appears in equation for: FitFun, InitEvenFF, InitKauffFF

Object Population

Contained in Object: Root

Containing Objects: SubString

Appears in equation for: Shift

ProbMut (P)	Parameter setting the probability for an element state modification by the agent. Appears in equation for: Mutation
Fitness (0)	Computes the fitness of the string, as obtained by the champions of all substrings. Appears in equation for: Shift
TestMe (0)	Provides the fitness of the string composed by the tentative bits of the agent and the current bits for the other sub-strings. Writes the global fitness in TestFit (P) and returns the expected local fitness of the agent Appears in equation for: Mutation, Shift

Table A5.4. (Cont.)

Object SubString
Contained in Object: Root → Population
Containing Objects: Agent, SSBit
Appears in equation for: Fitness, Shift

IdSS (P)	Parameter determining the identity of a specific substring. Appears in equation for: TestMe
Bidding (0)	The Substring scans all its agents finding the highest bid according to a specific criterion. It then returns the expected fitness for that bid. Appears in equation for: Fitness
ActLF (P)	Parameter measuring and saves the actual fitness for each substring in every simulation step. Appears in equation for: Fitness, Shift
Criterion (P)	Parameter determining how the substring selects the best-performing agent. If Criterion=1 (here), the agents are ranked according to the expected global fitness of their mutation. If Criterion=2, agents are evaluated based on the expected local fitness of their mutation. Appears in equation for: Bidding

Object Agent
Contained in Object: Root → Population → SubString
Containing Objects: ABit
Appears in equation for: Bidding, Shift

IdAgent (P)	Parameter allowing for an identification of all agents in the population. Appears in equation for: Bidding
Mutation (0)	Variable attempting a change of the current agent configuration (AValues). Appears in equation for: Bidding
Chosen (P)	The parameter chosen is set to 1 for the agent emerging as the champion of the bidding process in each simulation. That agent's configuration then enters as the new substring. Appears in equation for: Bidding, Fitness
ExpLocFit (P)	Parameter measuring the expected local fitness of a specific agent configuration in each simulation step, holding the remainder of the string constant. Appears in the equation for: Mutation, Bidding, Shift
ExpGlobFit (P)	ExpGlobFit measures the expected global (string) fitness of any agent's configuration in each simulation step, while holding the rest of the string constant. Appears in equation for: TestMe, Mutation, Bidding

Table A5.4. (Cont.)

Object ABit
Contained in Object: Root → Population → SubString → Agent
Appears in equation for: TestMe, Mutation, Fitness

IdAB (P)	Parameter identifying the bits under control of an agent. Appears in equation for: TestMe, Fitness
AValue (P)	Parameter corresponding to the fitness of the agent's current configuration (obtained in the last simulation step). Appears in equation for: Mutation, Fitness
TValue (P)	Parameter denoting the fitness of the configuration after mutation by the agent. The mutation (TValue) is only retained if its fitness is better than the previous one. Retention in turn depends on the selection mechanism underlying the TestMe function. Appears in equation for: TestMe, Mutation
FlagTest (P)	FlagTest decides whether an element is used in evaluating (testing) the fitness value of a mutation executed by an agent. If FlagTest=1, the respective element matters for the agent's evaluation of his strategy. If FlagTest=0, it does not. The expected local fitness (ExpLocFit) for an agent (determining his acceptance of a mutation) is calculated over the range of FlagTest. Appears in equation for: TestMe
FlagSString (P)	Parameter allocating each ABit to its corresponding SubString. Appears in equation for: TestMe, Mutation, Fitness

Object SSBit
Contained in Object: Root → Population → SubString
Appears in equation for: Fitness, Shift

IdSSB (P)	Parameter identifying the SSBits pertaining to a given SubString. Appears in equation for: TestMe, Fitness, Shift
Value (P)	Parameter representing the current element state [0;1]. Appears in equation for: TestMe, Fitness, Shift

Table A5.5. Variables and equations

Object Root

Variable Init; Used in: (never used) Using: CreateFitContrib N NumFF UseMem InitFF	Variable CreateFitContrib; Used in: Init Using: N

```
if(!strcmp(label,"Init"))
{ /* Equation setting the basic initializations.
Computed only once, it transforms itself into a pa-
rameter and is not computed again. Assigns the
Landscape object to a specific pointer to speed up
calls to the landscape */
p->cal("CreateFitContrib",0);
/ creates the memory location for the landscape
data
v[0]=V("N");
for(i=0; i<(int)v[0]; i++)
 RND>0.5?str[i]=0:str[i]=1;
v[2]=V("NumFF");
v[6]=V("UseMem");
if(v[6]==1)
{
try {
v[1]=pow(2,v[0]);
mydata=new double* [(long int)v[2]];
for(i=0; i<(int)v[2]; i++)
 {
  mydata[i]=NULL;
  mydata[i]=new double[(long int)v[1]];
  for(j=0; j<(long int)v[1]; j++)
   mydata[i][j]=-1;
 }
inmem=1;
 }
catch(...)
 {
 for(i=0; i<(int)v[2]; i++)
 {
 if(mydata[i]!=NULL)
  delete[] mydata[i];
 }
 delete[] mydata;
 plog("No memory\n");
 inmem=0;
 }
}
else  inmem=0;
CYCLE(cur1, "FF")  VS(cur1,"InitFF");
param=1;
res=0;
goto end;
}
```

```
if(!strcmp(label,"CreateFitContrib"))
{ /* Variable in the landscape that creates the
landscape values. Initialises the vectors used
for computing and storing the landscape val-
ues. */
v[0]=p->cal("N",0);
str=new unsigned int[(int)v[0]];
//temporary strings, used to store the binary
point
str2=new unsigned int[(int)v[0]]; //numbers
str3=new unsigned int[(int)v[0]]; //numbers
str4=new unsigned int[(int)v[0]]; //numbers
fc=new bit[(int)v[0]];
//memory structure used to store landscape.
See FitFun comments
for(v[1]=0; v[1]<v[0] ; v[1]++ )
{
 fc[(int)v[1]].l.l0=NULL;
 fc[(int)v[1]].l.l1=NULL;
}
res=1;
param=1;
goto end;
}
```

Table A5.5. (Cont.)

Variable Shift; Used in: Shift
Using: N NumFF PeriodShift Shift NumShiftBit FitFun IdBit FitContr appBit Fitness
TestMe ActLF ExpLocFit IdSSB Value

```
if(!strcmp(label,"Shift"))
{ /* Every PeriodShift steps the fitness contributions are shifted. That is, NumShiftBit of the fit-
ness contributions are changed and fitness of all agents, substrings, etc. has to be updated. */
v[0]=VL("Shift",1);
if(v[0]>1) END_EQUATION(v[0]-1);
v[0]=p->cal("N",0);
v[3]=VS(root,"NumFF");
v[7]=V("NumShiftBit");
CYCLE(cur, "FF")
 {
 CYCLES(cur,cur1, "Bit")
 WRITES(cur1,"appBit",1);
 for(v[8]=0; v[8]<v[7]; v[8]++)
 {
 cur1=RNDDRAWS(cur,"Bit","appBit");
 WRITES(cur1,"appBit",0);
 v[10]=VS(cur1,"IdBit")-1;
 shift(&fc[(int)v[10]].l, (int)v[3]);
 }
 }
CYCLE(cur, "Population")
 {
  v[11]=0;  v[13]=0;
  i=0;  CYCLES(cur, cur1, "SubString")
  {
  v[13]++; CYCLES(cur1, cur2, "Agent")
  {
   v[10]=VS_CHEAT(cur1->up,"TestMe", cur2);
   WRITES(cur2,"ExpLocFit",v[10]);
  }
  CYCLES(cur1, cur4, "SSBit")
   str[i++]=(int)VS(cur4,"Value");
  }
  v[5]=V("FitFun"); CYCLES(cur, cur1, "SubString")
  {
  v[8]=v[9]=0;  CYCLES(cur1, cur2, "SSBit")
  {
  v[6]=VS(cur2,"IdSSB");
  cur3=SEARCH_CND("IdBit",v[6]);
  v[7]=VS(cur3,"FitContr");
  v[8]+=v[7];  v[9]++;
  }
  WRITES(cur1,"ActLF",v[8]/v[9]);
  }
  WRITELS(cur,"Fitness",v[5],t);
 }
v[5]=V("PeriodShift"); res=v[5]; goto end;
}
```

Table A5.5. (Cont.)

Object FF

Variable InitEvenFF; Used in: InitFF	Variable InitFF; Used in: Init
Using: N EvenK ForeOverlap AftOverlap IdBit IdLink Exist	Using: InitEvenFF InitKauffFF TypeFF

```
FUNCTION("InitEvenFF") /* Initializes the FF       FUNCTION("InitFF")
object structure */                                /* Initializes the fitness functions */
v[0]=V("N");                                       v[0]=V("TypeFF");
v[3]=V("EvenK");                                   if(v[0]==1)
v[4]=V("ForeOverlap");                              v[1]=V("InitEvenFF");
v[5]=V("AftOverlap");                              else
cur=SEARCH("Bit");                                  v[1]=V("InitKauffFF");
ADDNOBJ("Bit", v[0]-1, cur);                       PARAMETER
v[1]=v[6]=1;                                        RESULT( 1)
v[7]=0;
CYCLE(cur, "Bit")
 {
 cur1=SEARCHS(cur,"Link");
 ADDNOBJS(cur,"Link", v[0]-1, cur1);
 v[2]=1;  CYCLES(cur, cur1, "Link")
  {
  v[8]=(v[6]-1)*v[3]+1;
  v[9]=v[6]*v[3];
  if(v[2]>=v[8] && v[2]<=v[9])      v[10]=1;
  else     v[10]=0;
  if(v[9]>v[0])
   v[9]=v[0];
  if(v[5]>0)
   {//consider the aftoverlap
   if(v[2]>v[9] && v[2]<=v[5]+v[9] )
    v[10]=1;
   if(v[9]==v[10] && v[2]<v[5])
    v[10]=1;
   }
  if(v[4]>0)
   {//consider the foreoverlap
   if(v[2]<v[8] && v[2]>=v[8]-v[4])
    v[10]=1;
   if(v[8]-v[4]<0 && v[2] > v[8]-v[4]+v[0])
    v[10]=1;
   }
  WRITES(cur1,"Exist",v[10]);
  WRITES(cur1,"IdLink",v[2]++);
  }
 v[7]++;
 if(v[7]==v[3])
  {
  v[7]=0;  v[6]++;
  }
 WRITES(cur,"IdBit",v[1]++);
 }
PARAMETER
RESULT(1 )
```

Table A5.5. (Cont.)

Variable FitFun; Used in: TestMe Fitness Shift
Using: N NumFF IdFF FitContr Exist

```
if(!strcmp(label,"FitFun")) { /* Computes the fitness of the binary point stored in "str". Over-
rules the standard Lsd automatic scheduling system. The equation is computed any time it is
requested and not only once in each time step.*/
last_update--;  if(c==NULL)
 {
 res=-1; goto end;
 }
v[0]=root->cal("N",0); v[13]=V("IdFF"); v[4]=VS(root,"NumFF");
if(inmem==1)
 {
 v[14]=bin2int(str,v[0]);
 if(mydata[(int)v[13]-1][(int)v[14]]!=-1) END_EQUATION(mydata[(int)v[13]-1][(int)v[14]]);
 }
v[1]=0; for(cur=SEARCH("Bit"),v[1]=0,i=0; i<(int)v[0]; i++, cur=go_brother(cur))
{//for each bit
v[2]=0; //assume that the fc exists
 for( cur1=SEARCHS(cur,"Link"), cl=&(fc[i].l), j=0; j<(int)v[0] ; j++, cur1=go_brother(cur1))
  {//for each link of the bit
  v[3]=VS(cur1,"Exist"); if(v[3]==1 && str[j]==1)
   {
   if(cl->l1!=NULL) cl=cl->l1; //the next link exists
   else
    {//create the next link cl->l1=new dynlink; cl->l1->l0=NULL; cl->l1->l1=NULL;
    if(j==(int)v[0]-1)
     {cl->l1->l=new leave; cl->l1->l->f=new double[(int)v[4]];   for(h=0; h<(int)v[4]; h++)
      cl->l1->l->f[h]=RND; //last link, create the fc
     }
    cl=cl->l1;
    }
   }
  else
   {//create the l0 dynlink
   if(cl->l0!=NULL)  cl=cl->l0; //the next link exists
   else
    {//create the next link
    cl->l0=new dynlink;  cl->l0->l0=NULL;  cl->l0->l1=NULL;
    if(j==(int)v[0]-1)
     {cl->l0->l=new leave; cl->l0->l->f=new double[(int)v[4]];  for(h=0; h<(int)v[4]; h++)
      cl->l0->l->f[h]=RND; //last link, create the fc
     }
    cl=cl->l0;
    }
   }
  }//end of for through links, therefore cl points to the last element, owning f
 v[1]+=cl->l->f[(int)v[13]-1]; //for the average
 WRITES(cur,"FitContr", cl->l->f[(int)v[13]-1]); //individual fc
 }//end of for through bits
res=v[1]/v[0];
if(inmem==1) mydata[(int)v[13]-1][(int)v[14]]=res; goto end;
}
```

Table A5.5. (Cont.)

Object Population

Variable Fitness; Used in: Shift Using: FitFun IdBit FitContr Bidding ActLF Chosen IdAB AValue FlagSString IdSSB Value	Variable TestMe; Used in: Mutation Shift Using: FitFun IdBit FitContr IdSS ExpGlobFit IdAB TValue FlagTest FlagSString IdSSB Value

```
/* Computes the fitness of the string, as obtained
by the champions of all substrings ensuring that
all sub strings have selected their best performing
agent. */
CYCLE(cur, "SubString")
 VS(cur,"Bidding");
i=0;
CYCLE(cur, "SubString")
 {
 cur1=SEARCH_CNDS(cur,"Chosen",1);
 CYCLES(cur1, cur2, "ABit")
  {
  if(VS(cur2, "FlagSString")==1)
   {
   v[3]=VS(cur2,"IdAB");
   cur3=SEARCH_CNDS(cur,"IdSSB",v[3]);
   v[4]=VS(cur2,"AValue");
   str[i++]=(int)v[4];
   WRITES(cur3,"Value",v[4]);
   }
  }
 }
v[5]=V("FitFun");
CYCLE(cur, "SubString")
 {
 v[8]=v[9]=0;
 CYCLES(cur, cur2, "SSBit")
  {
  v[6]=VS(cur2,"IdSSB");
  cur3=SEARCH_CND("IdBit",v[6]);
  v[7]=VS(cur3,"FitContr");
  v[8]+=v[7];
  v[9]++;
  }
 WRITES(cur,"ActLF",v[8]/v[9]);
 }
RESULT(v[5] )
```

```
/* Provides the fitness of the string com-
posed by the tentative bits of the agent and
the current bits for the other sub-strings.
Writes the global fitness in the parameter
TestFit and returns the local fitness of the
agent */
v[0]=VS(c->up,"IdSS");
i=0;
CYCLES(c,cur1, "ABit")
 {//place in str all bits marked by FlagSS-
tring, and those stored in the common string
for the rest
 if(VS(cur1, "FlagSString")==1)
  str[i++]=(int)VS(cur1, "TValue");
 else
  {
  v[10]=VS(cur1,"IdAB");
  cur=SEARCH_CND("IdSSB",v[10]);
  str[i++]=(int)VS(cur, "Value");
  }
 }
v[5]=V("FitFun"); //computes the fitness,
and set the fitness contributions
WRITES(c,"ExpGlobFit",v[5]);
v[2]=v[4]=0;
CYCLES(c, cur, "ABit")
 {
 if(VS(cur, "FlagTest")==1)
  {
  v[3]=VS(cur,"IdAB");
  cur1=SEARCH_CNDS(p->up,"IdBit",v[3]);
  v[2]+=VS(cur1,"FitContr");
  v[4]++;
  }
 }
RESULT(v[2]/v[4] )
```

Table A5.5. (Cont.)

Object SubString

Variable Bidding; Used in: Fitness

Using: Criterion IdAgent Mutation Chosen ExpLocFit ExpGlobFit

```
/* The SubString scans all its agent finding the highest bidding according to a specific 'criterion'.
It then returns the expected fitness. */
v[0]=-1; v[4]=VS(p,"Criterion");
CYCLE(cur, "Agent")
 {
 WRITES(cur,"Chosen",0);   VS(cur,"Mutation");
 if(v[4]==1)  v[1]=VS(cur,"ExpGlobFit");
 if(v[4]==2)  v[1]=VS(cur,"ExpLocFit");
 if(v[1]>v[0])
   {
   v[2]=VS(cur,"IdAgent");   cur2=cur;  v[0]=v[1];
   }
 }
WRITES(cur2,"Chosen",1);
RESULT(v[0] )
```

Object Agent

Variable Mutation; Used in: Bidding

Using: ProbMut TestMe ExpLocFit ExpGlobFit AValue TValue FlagSString

```
/* Attempts a mutation of the current AValues */
v[0]=VS(p->up->up,"ProbMut");  CYCLE(cur, "ABit")
 {
 if(VS(cur, "FlagSString")==1)
  {
  v[1]=VS(cur,"AValue");   WRITES(cur,"TValue",v[1]);
  }
 }
v[2]=VS(p->up->up,"TestMe"); v[4]=V("ExpGlobFit"); CYCLE(cur, "ABit")
 {
 if(VS(cur, "FlagSString")==1)
  { v[1]=VS(cur,"AValue"); if(RND<v[0]) v[1]==1?v[1]=0: v[1]=1;
WRITES(cur,"TValue",v[1]);}
 }
v[3]=VS(p->up->up,"TestMe"); v[5]=V("ExpGlobFit");
if(v[3]>v[2])
 {//TValues are better  CYCLE(cur, "ABit")
 {
 if(VS(cur, "FlagSString")==1)
 {v[1]=VS(cur,"TValue");   WRITES(cur,"AValue",v[1]); }
 }
 //leave the TestFit as it is  v[6]=1;  WRITE("ExpLocFit",v[3]);
 }
else
 {//AValues are better, replace the TestFit, modified by latest use of TestMe
 WRITE("ExpGlobFit",v[4]);   v[6]=0;  WRITE("ExpLocFit",v[2]);
 }
RESULT( v[6])
```

A6 Cluster adaptation to change – Results overview

Table A6.1. Co-ordination mechanisms in the different populations

Pop.	Label	Co-ordination mechanism
1	Individualistic	Agents select strategies that improve their own fitness, i.e. agents in the first group take the fitness of elements n_1-n_6 into account when evaluating their strategy, those in group two care for n_7-n_{12}, etc.
2	Collective	Agents select strategies that improve cluster fitness as a whole, i.e. all agents take the fitness of elements n_1-n_{24} into account when evaluating their strategies.
3	Alliance (1+2, 3+4)	Agents form alliances and select strategies that improve alliance fitness, i.e. agents in the first and second group take the fitness of elements n_1-n_{12} into account, agents in groups 3 and 4 consider n_{13}-n_{24}.
4	Alliance (1+3, 2+4)	Agents form alliances and select strategies that improve alliance fitness, i.e. agents in the first and third group consider the fitness of elements n_1-n_6 and n_{13}-n_{18}, agents in groups 2 and 4 consider n_7-n_{12}/n_{19}-n_{24}.
5	Alliance (1+4, 2+3)	Agents form alliances and select strategies that improve alliance fitness, i.e. agents in the first and fourth group consider the fitness of elements n_1-n_6 and n_{19}-n_{24}, agents in groups 2 and 3 consider n_7-n_{18}.
6	Leader firm (1)[130]	The cluster has a dominant agent in the first group. All other agents try to improve that agent's fitness alongside their own. Agents in groups 2-4 care for the elements n_1-n_6 alongside those under their control. The leader firm only cares for its own fitness, i.e. elements n_1-n_6.
7	Leader firm (2)	The cluster has a dominant agent in the second group. All other agents try to improve that agent's fitness alongside their own. Agents in groups 1, 3 and 4 care for the effects of their strategies on the fitness of elements n_7-n_{12} alongside those under their control. The leader firm only cares for its own fitness, i.e. elements n_7-n_{12}.
8	Leader firm (3)	The cluster has a dominant agent in the third group. All other agents try to improve that agent's fitness alongside their own. Agents in groups 1, 2 and 4 care for the effects of their strategies on the fitness of elements n_{13}-n_{18} alongside those under their control. The leader firm only cares for its own fitness, i.e. elements n_{13}-n_{18}.
9	Leader firm (4)	The cluster has a dominant agent in the fourth group. All other agents try to improve that agent's fitness alongside their own. Agents in groups 1-3 care for the effects of their strategies on the fitness of elements n_{19}-n_{24} alongside those under their control. The leader firm only cares for its own fitness, i.e. elements n_{19}-n_{24}.

---

[130] In leader firm scenario, the group containing the leading firm consists out of one instead of five agents.

Table A6.2. Landscape complexity and average system fitness (bifurcation)*

FL Complexity Pm**	K	C	Individu-alistic	Collective	Alliance (1+2 3+4)	Alliance (1+3 2+4)	Alliance (1+4 2+3=
4	2.33	0.67	0.72869	0.73685	0.72962	0.73722	0.72889
5	2.50	1.30	0.73078	0.73999	0.73293	0.73764	0.73189
7	3.17	2.33	0.73309	0.73921	0.73213	0.73643	0.72981
9	4.25	3.00	0.72303	0.73281	0.73332	0.73306	0.72314
8	3.67	3.33	0.73102	0.74306	0.74217	0.73802	0.73354
10	3.67	4.33	0.70568	0.72749	0.72368	0.72192	0.71543
11	3.92	5.33	0.69115	0.71906	0.70773	0.71699	0.70178
12	5.00	6.00	0.69580	0.71902	0.69791	0.71858	0.69605
13	4.58	6.50	0.66807	0.71536	0.68786	0.71217	0.68873
14	4.33	7.00	0.64741	0.71389	0.68223	0.71078	0.67646
15	4.25	7.50	0.62296	0.70868	0.66676	0.70737	0.67459
16	4.33	8.00	0.60742	0.70480	0.66699	0.69374	0.66373
17	4.58	8.50	0.60721	0.70822	0.66405	0.68571	0.66855
18	5.00	9.00	0.60406	0.70675	0.65926	0.66207	0.66709
19	4.58	10.50	0.58360	0.69924	0.64337	0.64880	0.65085
20	4.33	12.00	0.56127	0.68832	0.63241	0.63285	0.62572
21	4.25	13.50	0.54460	0.67797	0.61924	0.61550	0.61158
22	4.33	15.00	0.52972	0.67312	0.59629	0.59558	0.60021
23	4.58	16.50	0.51811	0.66556	0.58046	0.59267	0.58885
24	5.00	18.00	0.51334	0.66616	0.57491	0.57886	0.58349

FL Complexity Pm**	K	C	Leader firm (1)	Leader firm (2)	Leader firm (3)	Leader firm (4)
4	2.33	0.67	0.72612	0.72701	0.72839	0.72730
5	2.50	1.30	0.72984	0.72759	0.72399	0.72612
7	3.17	2.33	0.72360	0.72886	0.72554	0.72263
9	4.25	3.00	0.71693	0.71864	0.72542	0.72918
8	3.67	3.33	0.72593	0.73136	0.74200	0.72832
10	3.67	4.33	0.70987	0.71490	0.71875	0.70999
11	3.92	5.33	0.69918	0.70489	0.71020	0.69515
12	5.00	6.00	0.69651	0.70052	0.69839	0.70353
13	4.58	6.50	0.69885	0.68511	0.69876	0.69695
14	4.33	7.00	0.69448	0.67041	0.69482	0.69104
15	4.25	7.50	0.69037	0.65036	0.69061	0.68720
16	4.33	8.00	0.68346	0.62834	0.69021	0.68219
17	4.58	8.50	0.68373	0.61782	0.68310	0.68077
18	5.00	9.00	0.68020	0.60126	0.67967	0.67778
19	4.58	10.50	0.66269	0.60036	0.66135	0.66658
20	4.33	12.00	0.64201	0.58501	0.64470	0.65105
21	4.25	13.50	0.62414	0.59007	0.62933	0.62761
22	4.33	15.00	0.61010	0.57918	0.61324	0.60829
23	4.58	16.50	0.59812	0.57100	0.59370	0.59099
24	5.00	18.00	0.58167	0.58537	0.57664	0.58532

* All results reported correspond to averages over 100 simulations.
** Pm is the value of LSD parameter EvenK.

Table A6.3. Landscape complexity and standard deviation of system fitness (bifurcation)

FL Complexity			Individu-	Collective	Alliance	Alliance	Alliance
Pm**	K	C	alistic		(1+2 3+4)	(1+3 2+4)	(1+4 2+3)
4	2.33	0.67	0.01281	0.01029	0.00961	0.01366	0.01386
5	2.50	1.30	0.01900	0.01309	0.01600	0.01955	0.01643
7	3.17	2.33	0.02814	0.01766	0.02036	0.02755	0.02168
9	4.25	3.00	0.03981	0.02002	0.02723	0.03169	0.02381
8	3.67	3.33	0.03712	0.01984	0.02429	0.02582	0.01902
10	3.67	4.33	0.05133	0.02120	0.02656	0.03402	0.02735
11	3.92	5.33	0.05751	0.02281	0.02546	0.03980	0.03678
12	5.00	6.00	0.05448	0.02306	0.02284	0.04118	0.04381
13	4.58	6.50	0.06959	0.02410	0.02894	0.04145	0.04197
14	4.33	7.00	0.07870	0.02582	0.03663	0.03825	0.04128
15	4.25	7.50	0.08347	0.02581	0.04126	0.04567	0.04477
16	4.33	8.00	0.08023	0.02656	0.04855	0.04815	0.04439
17	4.58	8.50	0.07488	0.02540	0.05961	0.04869	0.04967
18	5.00	9.00	0.06845	0.02134	0.07044	0.05204	0.04892
19	4.58	10.50	0.06830	0.02568	0.07315	0.05546	0.05805
20	4.33	12.00	0.06822	0.02644	0.07831	0.06330	0.06303
21	4.25	13.50	0.06888	0.02654	0.08101	0.06949	0.06623
22	4.33	15.00	0.06795	0.02762	0.08378	0.07276	0.07407
23	4.58	16.50	0.06803	0.02926	0.08570	0.07679	0.07692
24	5.00	18.00	0.06790	0.02982	0.08893	0.08082	0.08055

FL Complexity			Leader firm	Leader firm	Leader firm	Leader firm
Pm**	K	C	(1)	(2)	(3)	(4)
4	2.33	0.67	0.01269	0.01306	0.01328	0.01302
5	2.50	1.30	0.01617	0.01835	0.01953	0.01718
7	3.17	2.33	0.01793	0.02378	0.02855	0.02327
9	4.25	3.00	0.02656	0.03212	0.04128	0.02675
8	3.67	3.33	0.02547	0.03849	0.03841	0.02437
10	3.67	4.33	0.03197	0.03819	0.04428	0.03414
11	3.92	5.33	0.03843	0.03902	0.05051	0.04817
12	5.00	6.00	0.04185	0.04159	0.05302	0.05346
13	4.58	6.50	0.04667	0.05237	0.06394	0.06146
14	4.33	7.00	0.05167	0.06129	0.06904	0.06823
15	4.25	7.50	0.05437	0.06737	0.07095	0.07368
16	4.33	8.00	0.05620	0.07085	0.07428	0.07150
17	4.58	8.50	0.05343	0.07008	0.07088	0.07325
18	5.00	9.00	0.05035	0.06842	0.06723	0.06907
19	4.58	10.50	0.05513	0.07252	0.07144	0.07514
20	4.33	12.00	0.06155	0.07729	0.08128	0.07673
21	4.25	13.50	0.06856	0.07793	0.08125	0.08192
22	4.33	15.00	0.07512	0.08051	0.08692	0.08533
23	4.58	16.50	0.07771	0.08278	0.08553	0.08683
24	5.00	18.00	0.07644	0.07966	0.08611	0.08825

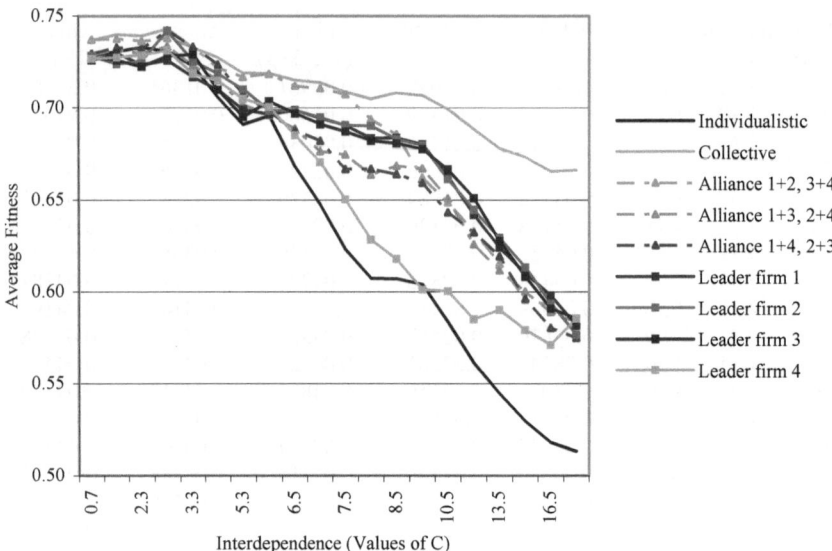

Fig. A6.1. Landscape complexity and adaptive performance (bifurcation)

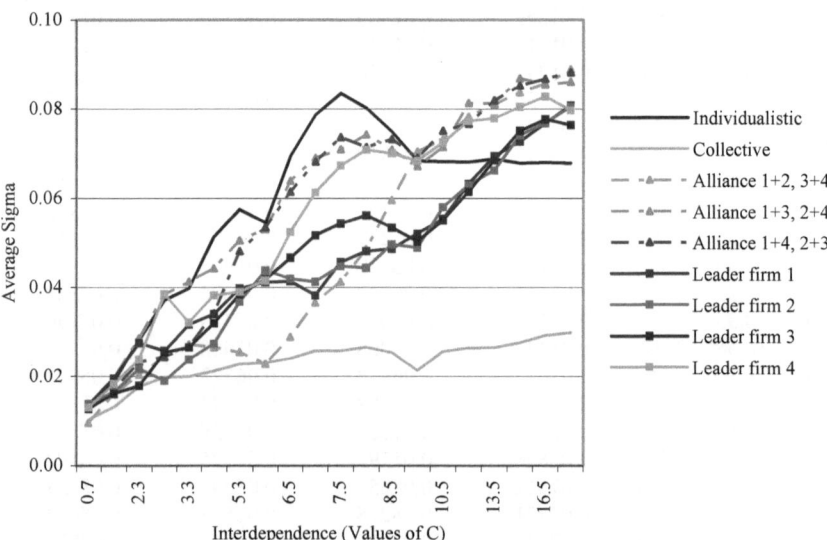

Fig. A6.2. Landscape complexity and stability in adaptation (bifurcation)

Table A6.4. Adaptation and group fitness (bifurcation)

FL Complexity			Individualistic			
Pm.	K	C	Group 1	Group 2	Group 3	Group 4
4	2.33	0.67	0.74021	0.72711	0.72341	0.72405
5	2.50	1.30	0.73321	0.73146	0.72241	0.73605
7	3.17	2.33	0.75869	0.73424	0.71778	0.72164
9	4.25	3.00	0.72625	0.70140	0.72601	0.77043
8	3.67	3.33	0.74243	0.71363	0.69287	0.74319
10	3.67	4.33	0.70960	0.69163	0.71242	0.70906
11	3.92	5.33	0.69894	0.68292	0.69303	0.68972
12	5.00	6.00	0.69998	0.69914	0.69194	0.69213
13	4.58	6.50	0.66022	0.65744	0.67323	0.68141
14	4.33	7.00	0.62246	0.62575	0.65667	0.68476
15	4.25	7.50	0.59270	0.59284	0.63012	0.67618
16	4.33	8.00	0.57094	0.57144	0.60452	0.68277
17	4.58	8.50	0.56044	0.55942	0.58263	0.72636
18	5.00	9.00	0.54747	0.54743	0.54709	0.77426
19	4.58	10.50	0.53904	0.53950	0.53913	0.71673
20	4.33	12.00	0.52996	0.53037	0.53060	0.65417
21	4.25	13.50	0.52528	0.52425	0.52458	0.60428
22	4.33	15.00	0.51791	0.51807	0.51898	0.56390
23	4.58	16.50	0.51502	0.51510	0.51583	0.52647
24	5.00	18.00	0.51343	0.51367	0.51355	0.51269

FL Complexity			Collective			
Pm.	K	C	Group 1	Group 2	Group 3	Group 4
4	2.33	0.67	0.75065	0.73130	0.73066	0.73480
5	2.50	1.30	0.74574	0.73890	0.72549	0.74982
7	3.17	2.33	0.75424	0.73891	0.73037	0.73332
9	4.25	3.00	0.74098	0.72820	0.73268	0.77036
8	3.67	3.33	0.74657	0.72040	0.73156	0.73269
10	3.67	4.33	0.73351	0.72922	0.70919	0.73804
11	3.92	5.33	0.72490	0.71338	0.71393	0.72402
12	5.00	6.00	0.72029	0.71858	0.71866	0.71856
13	4.58	6.50	0.70661	0.71302	0.72017	0.72162
14	4.33	7.00	0.70552	0.70590	0.72192	0.72221
15	4.25	7.50	0.68916	0.69571	0.70496	0.74488
16	4.33	8.00	0.69103	0.69169	0.70492	0.73155
17	4.58	8.50	0.67396	0.68799	0.71015	0.76077
18	5.00	9.00	0.69037	0.68189	0.68045	0.77428
19	4.58	10.50	0.68574	0.68600	0.67746	0.74776
20	4.33	12.00	0.66943	0.68094	0.67271	0.73022
21	4.25	13.50	0.66407	0.67500	0.67105	0.70176
22	4.33	15.00	0.67794	0.67685	0.66932	0.66838
23	4.58	16.50	0.67042	0.65476	0.66795	0.66912
24	5.00	18.00	0.66550	0.67176	0.66407	0.66333

Table A6.4. (Cont.)

FL Complexity			Alliance (1+2, 3+4)			
Pm.	*K*	*C*	Group 1	Group 2	Group 3	Group 4
4	*2.33*	*0.67*	0.74523	0.73523	0.73717	0.73127
5	*2.50*	*1.30*	0.74850	0.73223	0.71915	0.75068
7	*3.17*	*2.33*	0.75214	0.75074	0.70799	0.73484
9	*4.25*	*3.00*	0.71935	0.72026	0.74210	0.77035
8	*3.67*	*3.33*	0.74438	0.72758	0.71503	0.74525
10	*3.67*	*4.33*	0.72914	0.70469	0.72196	0.73191
11	*3.92*	*5.33*	0.72171	0.70427	0.72340	0.71858
12	*5.00*	*6.00*	0.71585	0.73105	0.71433	0.71308
13	*4.58*	*6.50*	0.70652	0.71297	0.70194	0.72726
14	*4.33*	*7.00*	0.70529	0.70245	0.70612	0.72924
15	*4.25*	*7.50*	0.69218	0.69300	0.71219	0.73212
16	*4.33*	*8.00*	0.66969	0.67354	0.70363	0.72809
17	*4.58*	*8.50*	0.64706	0.64790	0.68606	0.76181
18	*5.00*	*9.00*	0.62163	0.61432	0.63809	0.77426
19	*4.58*	*10.50*	0.61758	0.61934	0.62941	0.72888
20	*4.33*	*12.00*	0.60833	0.61544	0.62134	0.68629
21	*4.25*	*13.50*	0.60401	0.60286	0.60614	0.64899
22	*4.33*	*15.00*	0.58969	0.58810	0.58884	0.61568
23	*4.58*	*16.50*	0.59260	0.59398	0.59048	0.59360
24	*5.00*	*18.00*	0.58022	0.57709	0.58061	0.57754

FL Complexity			Alliance (1+3, 2+4)			
Pm.	*K*	*C*	Group 1	Group 2	Group 3	Group 4
4	*2.33*	*0.67*	0.73854	0.72860	0.72500	0.72343
5	*2.50*	*1.30*	0.73644	0.73169	0.72361	0.73582
7	*3.17*	*2.33*	0.75596	0.72827	0.71656	0.71846
9	*4.25*	*3.00*	0.74415	0.71393	0.69275	0.74175
8	*3.67*	*3.33*	0.72734	0.70676	0.72969	0.77039
10	*3.67*	*4.33*	0.71656	0.69184	0.73514	0.71816
11	*3.92*	*5.33*	0.70506	0.68893	0.71979	0.69335
12	*5.00*	*6.00*	0.69927	0.69607	0.69401	0.69485
13	*4.58*	*6.50*	0.67938	0.68273	0.69121	0.70159
14	*4.33*	*7.00*	0.65849	0.67530	0.66999	0.70206
15	*4.25*	*7.50*	0.65016	0.67296	0.66515	0.71008
16	*4.33*	*8.00*	0.63229	0.65737	0.64734	0.71792
17	*4.58*	*8.50*	0.63732	0.65480	0.63181	0.75027
18	*5.00*	*9.00*	0.62263	0.64910	0.62232	0.77433
19	*4.58*	*10.50*	0.62038	0.63685	0.61711	0.72904
20	*4.33*	*12.00*	0.60233	0.61015	0.60831	0.68207
21	*4.25*	*13.50*	0.60110	0.60293	0.59496	0.64735
22	*4.33*	*15.00*	0.59566	0.59535	0.59127	0.61858
23	*4.58*	*16.50*	0.58771	0.59050	0.58810	0.58912
24	*5.00*	*18.00*	0.58483	0.58667	0.58246	0.58002

Table A6.4. (Cont.)

FL Complexity			Alliance (1+4, 2+3)			
Pm.	*K*	*C*	Group 1	Group 2	Group 3	Group 4
4	*2.33*	*0.67*	0.74230	0.72643	0.72496	0.72479
5	*2.50*	*1.30*	0.73524	0.74149	0.71918	0.73582
7	*3.17*	*2.33*	0.75200	0.73137	0.72597	0.71917
9	*4.25*	*3.00*	0.75276	0.71482	0.71083	0.75488
8	*3.67*	*3.33*	0.74083	0.72447	0.73266	0.77073
10	*3.67*	*4.33*	0.72901	0.70828	0.73522	0.72221
11	*3.92*	*5.33*	0.71129	0.69376	0.71558	0.71029
12	*5.00*	*6.00*	0.70020	0.70322	0.69652	0.69168
13	*4.58*	*6.50*	0.68620	0.67830	0.68856	0.69839
14	*4.33*	*7.00*	0.68164	0.66484	0.67506	0.70737
15	*4.25*	*7.50*	0.66493	0.64193	0.65548	0.70468
16	*4.33*	*8.00*	0.66168	0.63944	0.65059	0.71624
17	*4.58*	*8.50*	0.64986	0.62357	0.63793	0.74485
18	*5.00*	*9.00*	0.63552	0.61276	0.61467	0.77408
19	*4.58*	*10.50*	0.62390	0.61334	0.60744	0.72880
20	*4.33*	*12.00*	0.62075	0.61371	0.61027	0.68490
21	*4.25*	*13.50*	0.60506	0.60690	0.60940	0.65562
22	*4.33*	*15.00*	0.58945	0.58870	0.59167	0.61532
23	*4.58*	*16.50*	0.58087	0.57958	0.57893	0.58246
24	*5.00*	*18.00*	0.57317	0.57527	0.57555	0.57566

FL Complexity			Leader firm (1)			
Pm.	*K*	*C*	Group 1	Group 2	Group 3	Group 4
4	*2.33*	*0.67*	0.75557	0.70339	0.72173	0.72380
5	*2.50*	*1.30*	0.75467	0.70688	0.72051	0.73730
7	*3.17*	*2.33*	0.75244	0.70850	0.71517	0.71828
9	*4.25*	*3.00*	0.76824	0.64808	0.70664	0.74475
8	*3.67*	*3.33*	0.76341	0.62881	0.74073	0.77076
10	*3.67*	*4.33*	0.76819	0.62604	0.73328	0.71198
11	*3.92*	*5.33*	0.77092	0.61780	0.71093	0.69709
12	*5.00*	*6.00*	0.77308	0.62873	0.69138	0.69286
13	*4.58*	*6.50*	0.78513	0.62785	0.67846	0.70398
14	*4.33*	*7.00*	0.76673	0.63273	0.65661	0.72184
15	*4.25*	*7.50*	0.75930	0.62928	0.63677	0.73612
16	*4.33*	*8.00*	0.75124	0.61840	0.62738	0.73681
17	*4.58*	*8.50*	0.74617	0.61687	0.61493	0.75696
18	*5.00*	*9.00*	0.72174	0.61286	0.61234	0.77386
19	*4.58*	*10.50*	0.71422	0.60917	0.60780	0.71955
20	*4.33*	*12.00*	0.69955	0.59995	0.59749	0.67106
21	*4.25*	*13.50*	0.67929	0.59022	0.59091	0.63615
22	*4.33*	*15.00*	0.67191	0.57849	0.58297	0.60703
23	*4.58*	*16.50*	0.66405	0.57510	0.58003	0.57332
24	*5.00*	*18.00*	0.63426	0.56735	0.56055	0.56452

Table A6.4. (Cont.)

FL Complexity			Leader firm (2)			
Pm.	K	C	Group 1	Group 2	Group 3	Group 4
4	2.33	0.67	0.72289	0.74039	0.72814	0.72213
5	2.50	1.30	0.71367	0.76312	0.68533	0.73383
7	3.17	2.33	0.73539	0.77025	0.66899	0.72754
9	4.25	3.00	0.71727	0.78514	0.64583	0.75343
8	3.67	3.33	0.70074	0.79883	0.69792	0.77051
10	3.67	4.33	0.66884	0.78849	0.70995	0.70773
11	3.92	5.33	0.64992	0.79709	0.70510	0.68871
12	5.00	6.00	0.63154	0.77814	0.69110	0.69276
13	4.58	6.50	0.62941	0.77646	0.67937	0.70982
14	4.33	7.00	0.62755	0.76379	0.65972	0.72822
15	4.25	7.50	0.63194	0.75843	0.63918	0.73287
16	4.33	8.00	0.62596	0.76255	0.63448	0.73785
17	4.58	8.50	0.61667	0.73498	0.62526	0.75547
18	5.00	9.00	0.60580	0.72910	0.60963	0.77414
19	4.58	10.50	0.60236	0.71007	0.61076	0.72220
20	4.33	12.00	0.60042	0.69890	0.60435	0.67513
21	4.25	13.50	0.59381	0.69201	0.59106	0.64045
22	4.33	15.00	0.58067	0.67348	0.58242	0.61638
23	4.58	16.50	0.57547	0.65329	0.57052	0.57553
24	5.00	18.00	0.56142	0.62512	0.55907	0.56095

FL Complexity			Leader firm (3)			
Pm.	K	C	Group 1	Group 2	Group 3	Group 4
4	2.33	0.67	0.73936	0.72430	0.74373	0.70183
5	2.50	1.30	0.73625	0.69923	0.76674	0.70226
7	3.17	2.33	0.75617	0.68958	0.77814	0.66662
9	4.25	3.00	0.75479	0.66395	0.77952	0.71846
8	3.67	3.33	0.74329	0.62617	0.77324	0.77057
10	3.67	4.33	0.73490	0.65279	0.77194	0.68032
11	3.92	5.33	0.71083	0.67321	0.76740	0.62917
12	5.00	6.00	0.69893	0.69570	0.77028	0.64920
13	4.58	6.50	0.68189	0.67798	0.78086	0.64708
14	4.33	7.00	0.66287	0.66684	0.77339	0.66106
15	4.25	7.50	0.64772	0.64375	0.77228	0.68506
16	4.33	8.00	0.63664	0.63936	0.75739	0.69537
17	4.58	8.50	0.61874	0.62489	0.74956	0.72990
18	5.00	9.00	0.60395	0.61327	0.71986	0.77405
19	4.58	10.50	0.60686	0.61252	0.72473	0.72220
20	4.33	12.00	0.60436	0.60783	0.71694	0.67507
21	4.25	13.50	0.59492	0.59803	0.69302	0.62449
22	4.33	15.00	0.57829	0.58612	0.66965	0.59908
23	4.58	16.50	0.56650	0.57685	0.65033	0.57029
24	5.00	18.00	0.57559	0.56172	0.64145	0.56253

Table A6.4. (Cont.)

FL Complexity			Leader firm (4)			
Pm.	K	C	Group 1	Group 2	Group 3	Group 4
4	2.33	0.67	0.73921	0.72565	0.69340	0.74978
5	2.50	1.30	0.73369	0.73175	0.68219	0.76274
7	3.17	2.33	0.75487	0.73726	0.66992	0.75340
9	4.25	3.00	0.74626	0.71706	0.64766	0.76358
8	3.67	3.33	0.72580	0.70754	0.72749	0.76463
10	3.67	4.33	0.72111	0.67651	0.71069	0.75131
11	3.92	5.33	0.71375	0.67694	0.67245	0.75643
12	5.00	6.00	0.69771	0.69972	0.63532	0.76932
13	4.58	6.50	0.67073	0.67339	0.62193	0.77440
14	4.33	7.00	0.65208	0.65546	0.61111	0.76298
15	4.25	7.50	0.61739	0.62067	0.59647	0.76691
16	4.33	8.00	0.59147	0.59269	0.57490	0.75429
17	4.58	8.50	0.57246	0.57020	0.56534	0.76328
18	5.00	9.00	0.54471	0.54651	0.54541	0.76841
19	4.58	10.50	0.56035	0.56083	0.56099	0.71926
20	4.33	12.00	0.55616	0.55767	0.55704	0.66917
21	4.25	13.50	0.56237	0.57578	0.56991	0.65220
22	4.33	15.00	0.56020	0.56315	0.56214	0.63124
23	4.58	16.50	0.55473	0.56044	0.55865	0.61018
24	5.00	18.00	0.56893	0.56383	0.57078	0.63795

Table A6.5. Landscape complexity and average system fitness (perturbation)*

FL Complexity			Individual-istic	Collective	Alliance (1+2; 3+4)	Alliance (1+3; 2+4)	Alliance (1+4; 2+3)
Pm.	K	C					
4	2.33	0.67	0.72800	0.73139	0.70885	0.73290	0.71974
5	2.50	1.33	0.74037	0.74282	0.72313	0.74412	0.72884
7	3.17	2.33	0.72851	0.73764	0.73210	0.72651	0.73307
9	4.25	3.00	0.72888	0.73747	0.71668	0.74171	0.72963
8	3.67	3.33	0.72974	0.73138	0.70820	0.73452	0.72252
10	3.67	4.33	0.72789	0.73240	0.70053	0.72967	0.72144
11	3.92	5.33	0.73003	0.72739	0.70351	0.73411	0.72070
12	5.00	6.00	0.72632	0.72429	0.70733	0.72650	0.72011
13	4.58	6.50	0.72992	0.73037	0.70681	0.72978	0.71688
14	4.33	7.00	0.72800	0.73139	0.70885	0.73290	0.71974
15	4.25	7.50	0.74037	0.74282	0.72313	0.74412	0.72884
16	4.33	8.00	0.73924	0.73537	0.72038	0.73798	0.72874
17	4.58	8.50	0.72974	0.73138	0.70820	0.73452	0.72252

* All results reported correspond to averages over 100 simulations.

Table A6.5. (Cont.)

FL Complexity			Leader firm	Leader firm	Leader firm	Leader firm
Pm.	*K*	*C*	(1)	(2)	(3)	(4)
4	2.33	0.67	0.71103	0.70012	0.69987	0.67411
5	2.50	1.33	0.73113	0.72480	0.72085	0.71963
7	3.17	2.33	0.72564	0.72528	0.72474	0.72644
9	4.25	3.00	0.73095	0.71555	0.69925	0.68963
8	3.67	3.33	0.72519	0.71150	0.70098	0.69733
10	3.67	4.33	0.70980	0.70694	0.70604	0.71182
11	3.92	5.33	0.72502	0.71044	0.70584	0.70321
12	5.00	6.00	0.71636	0.70996	0.70257	0.70202
13	4.58	6.50	0.72787	0.70903	0.70725	0.68840
14	4.33	7.00	0.71103	0.70012	0.69987	0.67411
15	4.25	7.50	0.73113	0.72480	0.72085	0.71963
16	4.33	8.00	0.73039	0.72771	0.72314	0.71656
17	4.58	8.50	0.72519	0.71150	0.70098	0.69733

Table A6.6. Landscape complexity and standard deviation of system fitness (perturbation)

FL Complexity			Individual-istic	Collective	Alliance (1+2; 3+4)	Alliance (1+3; 2+4)	Alliance (1+4; 2+3)
Pm.	*K*	*C*					
4	2.33	0.67	0.03646	0.03627	0.04705	0.04036	0.04359
5	2.50	1.33	0.03510	0.02870	0.03701	0.02612	0.02509
7	3.17	2.33	0.03645	0.03003	0.02937	0.03637	0.02922
9	4.25	3.00	0.03680	0.03310	0.04116	0.02730	0.03082
8	3.67	3.33	0.03829	0.03584	0.04601	0.04002	0.04227
10	3.67	4.33	0.03777	0.03643	0.04521	0.03376	0.03888
11	3.92	5.33	0.03615	0.03373	0.04158	0.02531	0.03217
12	5.00	6.00	0.03823	0.03619	0.04003	0.03250	0.02885
13	4.58	6.50	0.03579	0.03519	0.04212	0.04178	0.03714
14	4.33	7.00	0.03646	0.03627	0.04705	0.04036	0.04359
15	4.25	7.50	0.03510	0.02870	0.03701	0.02612	0.02509
16	4.33	8.00	0.03345	0.03143	0.03943	0.03081	0.02850
17	4.58	8.50	0.03829	0.03584	0.04601	0.04002	0.04227

FL Complexity			Leader firm	Leader firm	Leader firm	Leader firm
Pm.	*K*	*C*	(1)	(2)	(3)	(4)
4	2.33	0.67	0.05495	0.06321	0.05895	0.07622
5	2.50	1.33	0.02421	0.02417	0.02202	0.02455
7	3.17	2.33	0.03652	0.03266	0.02917	0.03136
9	4.25	3.00	0.03951	0.05225	0.05724	0.06644
8	3.67	3.33	0.04422	0.05515	0.05724	0.06722
10	3.67	4.33	0.04028	0.04574	0.04607	0.04345
11	3.92	5.33	0.03411	0.04403	0.04611	0.04456
12	5.00	6.00	0.03872	0.04341	0.04498	0.05152
13	4.58	6.50	0.04142	0.03987	0.04509	0.05572
14	4.33	7.00	0.05495	0.06321	0.05895	0.07622
15	4.25	7.50	0.02421	0.02417	0.02202	0.02455
16	4.33	8.00	0.03027	0.02993	0.02333	0.02663
17	4.58	8.50	0.04422	0.05515	0.05724	0.06722

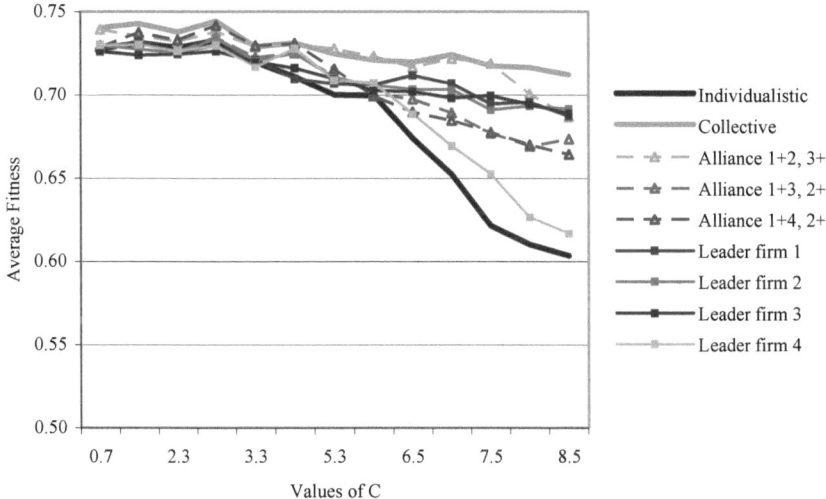

Fig. A6.3. Landscape complexity and adaptive performance (perturbation)

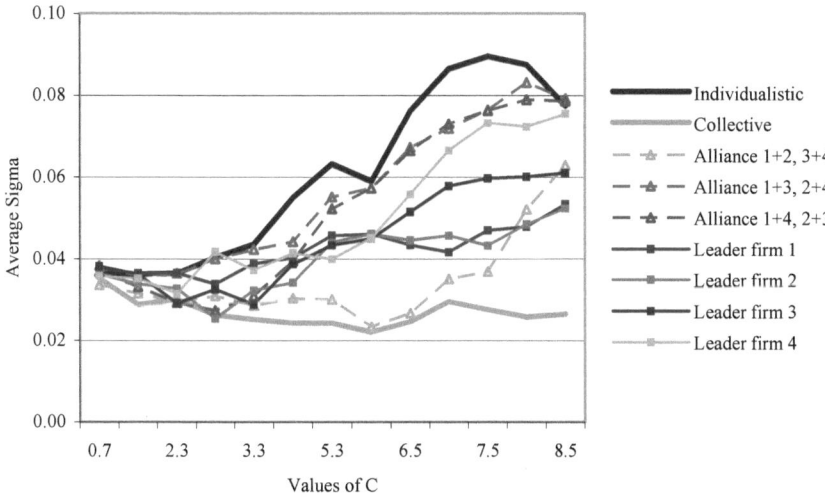

Fig. A6.4. Landscape complexity and stability of adaptation (perturbation)

Table A6.7. Adaptation and group fitness (perturbation)

FL Complexity			Individualistic			
Pm.	*K*	*C*	Group 1	Group 2	Group 3	Group 4
4	*2.33*	*0.67*	0.73260	0.72405	0.72638	0.72896
5	*2.50*	*1.30*	0.74478	0.71520	0.72891	0.73667
7	*3.17*	*2.33*	0.76350	0.72588	0.71509	0.70958
9	*4.25*	*3.00*	0.73746	0.70769	0.72627	0.76018
8	*3.67*	*3.33*	0.73468	0.70018	0.70641	0.73768
10	*3.67*	*4.33*	0.71987	0.70128	0.71098	0.71198
11	*3.92*	*5.33*	0.70343	0.69292	0.70395	0.70020
12	*5.00*	*6.00*	0.69244	0.69803	0.70477	0.70424
13	*4.58*	*6.50*	0.66335	0.66760	0.68153	0.68398
14	*4.33*	*7.00*	0.62990	0.63066	0.66505	0.68425
15	*4.25*	*7.50*	0.59312	0.59531	0.62838	0.66819
16	*4.33*	*8.00*	0.57506	0.57556	0.60683	0.68326
17	*4.58*	*8.50*	0.55471	0.55615	0.57738	0.72591

FL Complexity			Collective			
Pm.	*K*	*C*	Group 1	Group 2	Group 3	Group 4
4	*2.33*	*0.67*	0.74116	0.73787	0.74334	0.73909
5	*2.50*	*1.30*	0.76510	0.71915	0.73425	0.75279
7	*3.17*	*2.33*	0.76069	0.72125	0.75740	0.71124
9	*4.25*	*3.00*	0.74697	0.73755	0.73161	0.76034
8	*3.67*	*3.33*	0.73003	0.73823	0.72404	0.72308
10	*3.67*	*4.33*	0.74082	0.72419	0.70858	0.75093
11	*3.92*	*5.33*	0.73201	0.72381	0.73837	0.70502
12	*5.00*	*6.00*	0.72450	0.71857	0.72861	0.71170
13	*4.58*	*6.50*	0.70299	0.72340	0.72857	0.72355
14	*4.33*	*7.00*	0.70947	0.72941	0.73246	0.72546
15	*4.25*	*7.50*	0.71028	0.72530	0.71918	0.71484
16	*4.33*	*8.00*	0.70510	0.69627	0.71940	0.74504
17	*4.58*	*8.50*	0.68631	0.69272	0.70135	0.76789

FL Complexity			Alliance (1+2, 3+4)			
Pm.	*K*	*C*	Group 1	Group 2	Group 3	Group 4
4	*2.33*	*0.67*	0.74230	0.73611	0.74283	0.73571
5	*2.50*	*1.30*	0.76101	0.70090	0.72843	0.75115
7	*3.17*	*2.33*	0.74651	0.73903	0.72003	0.72283
9	*4.25*	*3.00*	0.73427	0.71515	0.74226	0.76023
8	*3.67*	*3.33*	0.72940	0.72466	0.72778	0.73312
10	*3.67*	*4.33*	0.73822	0.71852	0.71807	0.74675
11	*3.92*	*5.33*	0.73870	0.72129	0.72821	0.72265
12	*5.00*	*6.00*	0.71451	0.73123	0.70226	0.74455
13	*4.58*	*6.50*	0.71556	0.71943	0.70356	0.72770
14	*4.33*	*7.00*	0.71282	0.71091	0.73714	0.72646
15	*4.25*	*7.50*	0.70621	0.70675	0.72580	0.73716
16	*4.33*	*8.00*	0.67703	0.67533	0.69619	0.75313
17	*4.58*	*8.50*	0.64991	0.65140	0.68912	0.75576

Table A6.7. (Cont.)

FL Complexity			Alliance (1+3, 2+4)			
Pm.	K	C	Group 1	Group 2	Group 3	Group 4
4	2.33	0.67	0.73288	0.72517	0.72866	0.73226
5	2.50	1.30	0.74584	0.71201	0.73028	0.73738
7	3.17	2.33	0.76134	0.71861	0.71510	0.71098
9	4.25	3.00	0.73935	0.70934	0.72930	0.76006
8	3.67	3.33	0.74041	0.70353	0.70670	0.73943
10	3.67	4.33	0.73140	0.69897	0.73938	0.73101
11	3.92	5.33	0.70884	0.70502	0.72786	0.70427
12	5.00	6.00	0.69848	0.70127	0.70168	0.70248
13	4.58	6.50	0.68911	0.70572	0.69339	0.70112
14	4.33	7.00	0.66623	0.68916	0.68608	0.71463
15	4.25	7.50	0.66103	0.67478	0.65903	0.71322
16	4.33	8.00	0.64535	0.66860	0.64477	0.71793
17	4.58	8.50	0.63583	0.66857	0.64042	0.74887

FL Complexity			Alliance (1+4, 2+3)			
Pm.	K	C	Group 1	Group 2	Group 3	Group 4
4	2.33	0.67	0.73305	0.72283	0.72849	0.73116
5	2.50	1.30	0.74657	0.72653	0.74060	0.73617
7	3.17	2.33	0.76451	0.71334	0.74047	0.71394
9	4.25	3.00	0.75129	0.72643	0.72877	0.76037
8	3.67	3.33	0.75282	0.71710	0.70484	0.74375
10	3.67	4.33	0.74203	0.71380	0.73952	0.72847
11	3.92	5.33	0.72123	0.71554	0.71597	0.70948
12	5.00	6.00	0.69827	0.69821	0.70292	0.69762
13	4.58	6.50	0.68364	0.68121	0.69280	0.70088
14	4.33	7.00	0.68145	0.66509	0.68013	0.71192
15	4.25	7.50	0.67717	0.66257	0.65887	0.71148
16	4.33	8.00	0.66852	0.63595	0.65823	0.71794
17	4.58	8.50	0.65096	0.62823	0.63236	0.74559

FL Complexity			Leader firm (1)			
Pm.	K	C	Group 1	Group 2	Group 3	Group 4
4	2.33	0.67	0.75437	0.69146	0.72751	0.73820
5	2.50	1.30	0.76691	0.69635	0.73072	0.73564
7	3.17	2.33	0.77150	0.69986	0.72211	0.70911
9	4.25	3.00	0.79331	0.63369	0.73143	0.76026
8	3.67	3.33	0.77015	0.66288	0.71403	0.73870
10	3.67	4.33	0.76852	0.62057	0.72716	0.72296
11	3.92	5.33	0.78296	0.61863	0.72014	0.70604
12	5.00	6.00	0.77971	0.64389	0.70157	0.69899
13	4.58	6.50	0.78815	0.65001	0.68500	0.72412
14	4.33	7.00	0.78594	0.63884	0.66112	0.74183
15	4.25	7.50	0.77151	0.64282	0.62794	0.73643
16	4.33	8.00	0.77650	0.63831	0.62594	0.74198
17	4.58	8.50	0.75974	0.61337	0.61492	0.76084

Table A6.7. (Cont.)

FL Complexity			Leader firm (2)			
Pm.	K	C	Group 1	Group 2	Group 3	Group 4
4	2.33	0.67	0.70561	0.75298	0.72749	0.73403
5	2.50	1.30	0.72011	0.75071	0.70275	0.73600
7	3.17	2.33	0.74182	0.77700	0.67084	0.71147
9	4.25	3.00	0.68288	0.80065	0.69270	0.76022
8	3.67	3.33	0.70794	0.78352	0.64841	0.74293
10	3.67	4.33	0.68632	0.79758	0.70947	0.70670
11	3.92	5.33	0.64272	0.79812	0.70763	0.69330
12	5.00	6.00	0.64450	0.77581	0.70106	0.70198
13	4.58	6.50	0.63956	0.77161	0.68423	0.71745
14	4.33	7.00	0.63532	0.77024	0.66869	0.73848
15	4.25	7.50	0.63287	0.76819	0.62713	0.73548
16	4.33	8.00	0.63169	0.77141	0.62531	0.74432
17	4.58	8.50	0.62843	0.76170	0.61843	0.75748

FL Complexity			Leader firm (3)			
Pm.	K	C	Group 1	Group 2	Group 3	Group 4
4	2.33	0.67	0.73309	0.72467	0.75326	0.69427
5	2.50	1.30	0.74406	0.68300	0.77607	0.69402
7	3.17	2.33	0.76285	0.68603	0.79434	0.65572
9	4.25	3.00	0.75117	0.63252	0.76224	0.76007
8	3.67	3.33	0.74406	0.62937	0.79767	0.70935
10	3.67	4.33	0.74378	0.65502	0.77759	0.68907
11	3.92	5.33	0.71643	0.68557	0.78849	0.64936
12	5.00	6.00	0.69507	0.70060	0.78203	0.63258
13	4.58	6.50	0.68218	0.68854	0.79896	0.63838
14	4.33	7.00	0.67734	0.66635	0.78159	0.66764
15	4.25	7.50	0.67372	0.66558	0.77333	0.68558
16	4.33	8.00	0.64325	0.65638	0.77167	0.70613
17	4.58	8.50	0.63066	0.62873	0.76528	0.73128

FL Complexity			Leader firm (4)			
Pm.	K	C	Group 1	Group 2	Group 3	Group 4
4	2.33	0.67	0.73244	0.72349	0.70995	0.75379
5	2.50	1.30	0.74348	0.71886	0.69197	0.76718
7	3.17	2.33	0.76144	0.73007	0.66851	0.74573
9	4.25	3.00	0.73508	0.70348	0.72307	0.75750
8	3.67	3.33	0.73960	0.71483	0.64254	0.77055
10	3.67	4.33	0.73885	0.69550	0.70997	0.76715
11	3.92	5.33	0.71629	0.67908	0.67950	0.76126
12	5.00	6.00	0.70008	0.70375	0.65033	0.77485
13	4.58	6.50	0.67509	0.67885	0.62137	0.77829
14	4.33	7.00	0.64584	0.64710	0.61128	0.77266
15	4.25	7.50	0.62347	0.62621	0.59629	0.76363
16	4.33	8.00	0.58909	0.58984	0.56278	0.76380
17	4.58	8.50	0.56840	0.56903	0.55750	0.77122

Table A6.8. Payoff for individualistic and egoistic groups (bifurcation)

FL Complexity			*Individualistic*			
Pm.	*K*	*C*	*Group 1*	*Group 2*	*Group 3*	*Group 4*
4	*2.33*	*0.67*	*0.74021*	*0.72711*	*0.72341*	*0.72405*
5	*2.50*	*1.30*	*0.73321*	*0.73146*	*0.72241*	*0.73605*
7	*3.17*	*2.33*	*0.75869*	*0.73424*	*0.71778*	*0.72164*
9	*4.25*	*3.00*	*0.74243*	*0.71363*	*0.69287*	*0.74319*
8	*3.67*	*3.33*	*0.72625*	*0.70140*	*0.72601*	*0.77043*
10	*3.67*	*4.33*	*0.70960*	*0.69163*	*0.71242*	*0.70906*
11	*3.92*	*5.33*	*0.69894*	*0.68292*	*0.69303*	*0.68972*
12	*5.00*	*6.00*	*0.69998*	*0.69914*	*0.69194*	*0.69213*

FL Complexity			Collective			
Pm.	*K*	*C*	**Group 1**	**Group 2**	**Group 3**	**Group 4**
4	*2.33*	*0.67*	0.75065	0.73130	0.73066	0.73480
5	*2.50*	*1.30*	0.74574	0.73890	0.72549	0.74982
7	*3.17*	*2.33*	0.75424	0.73891	0.73037	0.73332
9	*4.25*	*3.00*	0.74657	0.72040	0.73156	0.73269
8	*3.67*	*3.33*	0.74098	0.72820	0.73268	0.77036
10	*3.67*	*4.33*	0.73351	0.72922	0.70919	0.73804
11	*3.92*	*5.33*	0.72490	0.71338	0.71393	0.72402
12	*5.00*	*6.00*	0.72029	0.71858	0.71866	0.71856

FL Complexity			Egoistic group 1			
Pm.	*K*	*C*	**Group 1**	*Group 2*	Group 3	Group 4
4	*2.33*	*0.67*	**0.76234***	*0.72064*	0.73520	0.73847
5	*2.50*	*1.30*	**0.75680**	*0.72277*	0.71457	0.74804
7	*3.17*	*2.33*	**0.76393**	*0.70712*	0.73040	0.72414
9	*4.25*	*3.00*	**0.76888**	*0.66695*	0.72879	0.74553
8	*3.67*	*3.33*	**0.76599**	*0.67863*	0.73847	0.76529
10	*3.67*	*4.33*	**0.76538**	*0.65358*	0.71310	0.72564
11	*3.92*	*5.33*	**0.77342**	*0.64126*	0.71451	0.71415
12	*5.00*	*6.00*	**0.77657**	*0.66662*	0.72230	0.72236

FL Complexity			Egoistic group 2			
Pm.	*K*	*C*	*Group 1*	**Group 2**	Group 3	Group 4
4	*2.33*	*0.67*	*0.72445*	**0.75436**	0.73313	0.74035
5	*2.50*	*1.30*	*0.72182*	**0.75981**	0.70461	0.74070
7	*3.17*	*2.33*	*0.74782*	**0.77278**	0.67850	0.73043
9	*4.25*	*3.00*	*0.72499*	**0.77032**	0.67194	0.74122
8	*3.67*	*3.33*	*0.69737*	**0.79052**	0.69367	0.76525
10	*3.67*	*4.33*	*0.67611*	**0.78425**	0.69653	0.72683
11	*3.92*	*5.33*	*0.65692*	**0.77378**	0.71610	0.70240
12	*5.00*	*6.00*	*0.65694*	**0.77390**	0.71716	0.72104

* Numbers in **bold** represent the fitness of the egoistic agent group, numbers in *italics* that of its neighbour. Comparing these numbers to the benchmark scenarios of *individualistic* and collective modes of co-ordination yields the aforementioned prisoner's dilemma payoff structure where: **Egoistic group**>Collective>*Individualistic*>*Egoistic neighbour*.

Table A6.8. (Cont.)

FL Complexity			*Individualistic*			
Pm.	*K*	*C*	*Group 1*	*Group 2*	*Group 3*	*Group 4*
4	*2.33*	*0.67*	*0.74021*	*0.72711*	*0.72341*	*0.72405*
5	*2.50*	*1.30*	*0.73321*	*0.73146*	*0.72241*	*0.73605*
7	*3.17*	*2.33*	*0.75869*	*0.73424*	*0.71778*	*0.72164*
9	*4.25*	*3.00*	*0.74243*	*0.71363*	*0.69287*	*0.74319*
8	*3.67*	*3.33*	*0.72625*	*0.70140*	*0.72601*	*0.77043*
10	*3.67*	*4.33*	*0.70960*	*0.69163*	*0.71242*	*0.70906*
11	*3.92*	*5.33*	*0.69894*	*0.68292*	*0.69303*	*0.68972*
12	*5.00*	*6.00*	*0.69998*	*0.69914*	*0.69194*	*0.69213*

FL Complexity			**Collective**			
Pm.	*K*	*C*	**Group 1**	**Group 2**	**Group 3**	**Group 4**
4	*2.33*	*0.67*	**0.75065**	**0.73130**	**0.73066**	**0.73480**
5	*2.50*	*1.30*	**0.74574**	**0.73890**	**0.72549**	**0.74982**
7	*3.17*	*2.33*	**0.75424**	**0.73891**	**0.73037**	**0.73332**
9	*4.25*	*3.00*	**0.74657**	**0.72040**	**0.73156**	**0.73269**
8	*3.67*	*3.33*	**0.74098**	**0.72820**	**0.73268**	**0.77036**
10	*3.67*	*4.33*	**0.73351**	**0.72922**	**0.70919**	**0.73804**
11	*3.92*	*5.33*	**0.72490**	**0.71338**	**0.71393**	**0.72402**
12	*5.00*	*6.00*	**0.72029**	**0.71858**	**0.71866**	**0.71856**

FL Complexity			Egoistic group 3			
Pm.	*K*	*C*	Group 1	Group 2	**Group 3**	*Group 4*
4	*2.33*	*0.67*	0.74209	0.74722	**0.74439**	*0.71495*
5	*2.50*	*1.30*	0.74960	0.70120	**0.76570**	*0.70662*
7	*3.17*	*2.33*	0.75059	0.71582	**0.78545**	*0.68549*
9	*4.25*	*3.00*	0.74056	0.67850	**0.77368**	*0.73207*
8	*3.67*	*3.33*	0.74076	0.65827	**0.77342**	*0.76540*
10	*3.67*	*4.33*	0.72620	0.68165	**0.76635**	*0.67713*
11	*3.92*	*5.33*	0.72258	0.71732	**0.76987**	*0.65547*
12	*5.00*	*6.00*	0.71279	0.72224	**0.77171**	*0.66743*

FL Complexity			Egoistic group 4			
Pm.	*K*	*C*	Group 1	Group 2	*Group 3*	**Group 4**
4	*2.33*	*0.67*	0.74995	0.73952	*0.71053*	**0.75444**
5	*2.50*	*1.30*	0.75027	0.73103	*0.69040*	**0.76313**
7	*3.17*	*2.33*	0.75001	0.74638	*0.68507*	**0.75684**
9	*4.25*	*3.00*	0.75145	0.71921	*0.67591*	**0.77285**
8	*3.67*	*3.33*	0.72241	0.73323	*0.73553*	**0.76531**
10	*3.67*	*4.33*	0.72285	0.71307	*0.70414*	**0.74924**
11	*3.92*	*5.33*	0.70792	0.72226	*0.67161*	**0.76315**
12	*5.00*	*6.00*	0.70295	0.71639	*0.65000*	**0.77480**

Table A6.9. Payoff for individualistic and egoistic groups (perturbation)

FL Complexity			*Individualistic*			
Pm.	*K*	*C*	*Group 1*	*Group 2*	*Group 3*	*Group 4*
4	*2.33*	*0.67*	*0.73260*	*0.72405*	*0.72638*	*0.72896*
5	*2.50*	*1.30*	*0.74478*	*0.71520*	*0.72891*	*0.73667*
7	*3.17*	*2.33*	*0.73617*	*0.70862*	*0.69087*	*0.69977*
9	*4.25*	*3.00*	*0.73468*	*0.70018*	*0.70641*	*0.73768*
8	*3.67*	*3.33*	*0.73746*	*0.70769*	*0.72627*	*0.76018*
10	*3.67*	*4.33*	*0.71987*	*0.70128*	*0.71098*	*0.71198*
11	*3.92*	*5.33*	*0.70343*	*0.69292*	*0.70395*	*0.70020*
12	*5.00*	*6.00*	*0.69244*	*0.69803*	*0.70477*	*0.70424*

FL Complexity			Collective			
Pm.	*K*	*C*	**Group 1**	**Group 2**	**Group 3**	**Group 4**
4	*2.33*	*0.67*	0.74116	0.73787	0.74334	0.73909
5	*2.50*	*1.30*	0.76510	0.71915	0.73425	0.75279
7	*3.17*	*2.33*	0.74073	0.72266	0.71482	0.71432
9	*4.25*	*3.00*	0.73003	0.73823	0.72404	0.72308
8	*3.67*	*3.33*	0.74697	0.73755	0.73161	0.76034
10	*3.67*	*4.33*	0.74082	0.72419	0.70858	0.75093
11	*3.92*	*5.33*	0.73201	0.72381	0.73837	0.70502
12	*5.00*	*6.00*	0.72450	0.71857	0.72861	0.71170

FL Complexity			Egoistic group 1			
Pm.	*K*	*C*	**Group 1**	*Group 2*	Group 3	Group 4
4	*2.33*	*0.67*	**0.75342***	*0.72324*	0.74486	0.74124
5	*2.50*	*1.30*	**0.75539**	*0.70813*	0.73180	0.75285
7	*3.17*	*2.33*	**0.76816**	*0.69622*	0.73068	0.73051
9	*4.25*	*3.00*	**0.75621**	*0.66920*	0.73224	0.73881
8	*3.67*	*3.33*	**0.77977**	*0.66032*	0.73019	0.76054
10	*3.67*	*4.33*	**0.78230**	*0.65971*	0.71908	0.74948
11	*3.92*	*5.33*	**0.78297**	*0.64755*	0.71914	0.70989
12	*5.00*	*6.00*	**0.78019**	*0.65581*	0.72092	0.72627

FL Complexity			Egoistic group 2			
Pm.	*K*	*C*	*Group 1*	**Group 2**	Group 3	Group 4
4	*2.33*	*0.67*	*0.70013*	**0.76893**	0.75643	0.73382
5	*2.50*	*1.30*	*0.70957*	**0.77020**	0.67533	0.75947
7	*3.17*	*2.33*	*0.74181*	**0.78014**	0.67406	0.72524
9	*4.25*	*3.00*	*0.71910*	**0.78972**	0.66924	0.74497
8	*3.67*	*3.33*	*0.70407*	**0.79572**	0.70797	0.76036
10	*3.67*	*4.33*	*0.68534*	**0.79056**	0.70436	0.74703
11	*3.92*	*5.33*	*0.66237*	**0.78446**	0.71385	0.70268
12	*5.00*	*6.00*	*0.65522*	**0.78030**	0.71133	0.72401

* Numbers in **bold** represent the fitness of the egoistic agent group, numbers in *italics* that of its neighbour. Comparing these numbers to the benchmark scenarios of *individualistic* and collective modes of co-ordination yields the aforementioned prisoner's dilemma payoff structure where: **Egoistic group**>Collective>*Individualistic*>*Egoistic neighbour.*

Table A6.9. (Cont.)

FL Complexity			*Individualistic*			
Pm.	*K*	*C*	*Group 1*	*Group 2*	*Group 3*	*Group 4*
4	*2.33*	*0.67*	*0.73260*	*0.72405*	*0.72638*	*0.72896*
5	*2.50*	*1.30*	*0.74478*	*0.71520*	*0.72891*	*0.73667*
7	*3.17*	*2.33*	*0.73617*	*0.70862*	*0.69087*	*0.69977*
9	*4.25*	*3.00*	*0.73468*	*0.70018*	*0.70641*	*0.73768*
8	*3.67*	*3.33*	*0.73746*	*0.70769*	*0.72627*	*0.76018*
10	*3.67*	*4.33*	*0.71987*	*0.70128*	*0.71098*	*0.71198*
11	*3.92*	*5.33*	*0.70343*	*0.69292*	*0.70395*	*0.70020*
12	*5.00*	*6.00*	*0.69244*	*0.69803*	*0.70477*	*0.70424*

FL Complexity			**Collective**			
Pm.	*K*	*C*	**Group 1**	**Group 2**	**Group 3**	**Group 4**
4	*2.33*	*0.67*	**0.74116**	**0.73787**	**0.74334**	**0.73909**
5	*2.50*	*1.30*	**0.76510**	**0.71915**	**0.73425**	**0.75279**
7	*3.17*	*2.33*	**0.74073**	**0.72266**	**0.71482**	**0.71432**
9	*4.25*	*3.00*	**0.73003**	**0.73823**	**0.72404**	**0.72308**
8	*3.67*	*3.33*	**0.74697**	**0.73755**	**0.73161**	**0.76034**
10	*3.67*	*4.33*	**0.74082**	**0.72419**	**0.70858**	**0.75093**
11	*3.92*	*5.33*	**0.73201**	**0.72381**	**0.73837**	**0.70502**
12	*5.00*	*6.00*	**0.72450**	**0.71857**	**0.72861**	**0.71170**

FL Complexity			Egoistic group 3			
Pm.	*K*	*C*	Group 1	Group 2	**Group 3**	*Group 4*
4	*2.33*	*0.67*	0.72860	0.74718	**0.75974**	*0.71431*
5	*2.50*	*1.30*	0.75139	0.68587	**0.76440**	*0.71887*
7	*3.17*	*2.33*	0.75495	0.72070	**0.77503**	*0.67120*
9	*4.25*	*3.00*	0.73662	0.66662	**0.78608**	*0.71622*
8	*3.67*	*3.33*	0.74388	0.66368	**0.76473**	*0.76036*
10	*3.67*	*4.33*	0.72998	0.68700	**0.77797**	*0.70311*
11	*3.92*	*5.33*	0.72715	0.70629	**0.77244**	*0.64337*
12	*5.00*	*6.00*	0.72772	0.72892	**0.78621**	*0.65469*

FL Complexity			Egoistic group 4			
Pm.	*K*	*C*	Group 1	Group 2	*Group 3*	**Group 4**
4	*2.33*	*0.67*	0.73103	0.74793	*0.72123*	**0.74839**
5	*2.50*	*1.30*	0.75630	0.71589	*0.71952*	**0.76104**
7	*3.17*	*2.33*	0.75970	0.74876	*0.68045*	**0.75911**
9	*4.25*	*3.00*	0.72308	0.74081	*0.67640*	**0.75567**
8	*3.67*	*3.33*	0.72850	0.75610	*0.72706*	**0.73372**
10	*3.67*	*4.33*	0.73716	0.72078	*0.70477*	**0.77092**
11	*3.92*	*5.33*	0.72852	0.71155	*0.68596*	**0.74120**
12	*5.00*	*6.00*	0.72143	0.73018	*0.65985*	**0.77701**

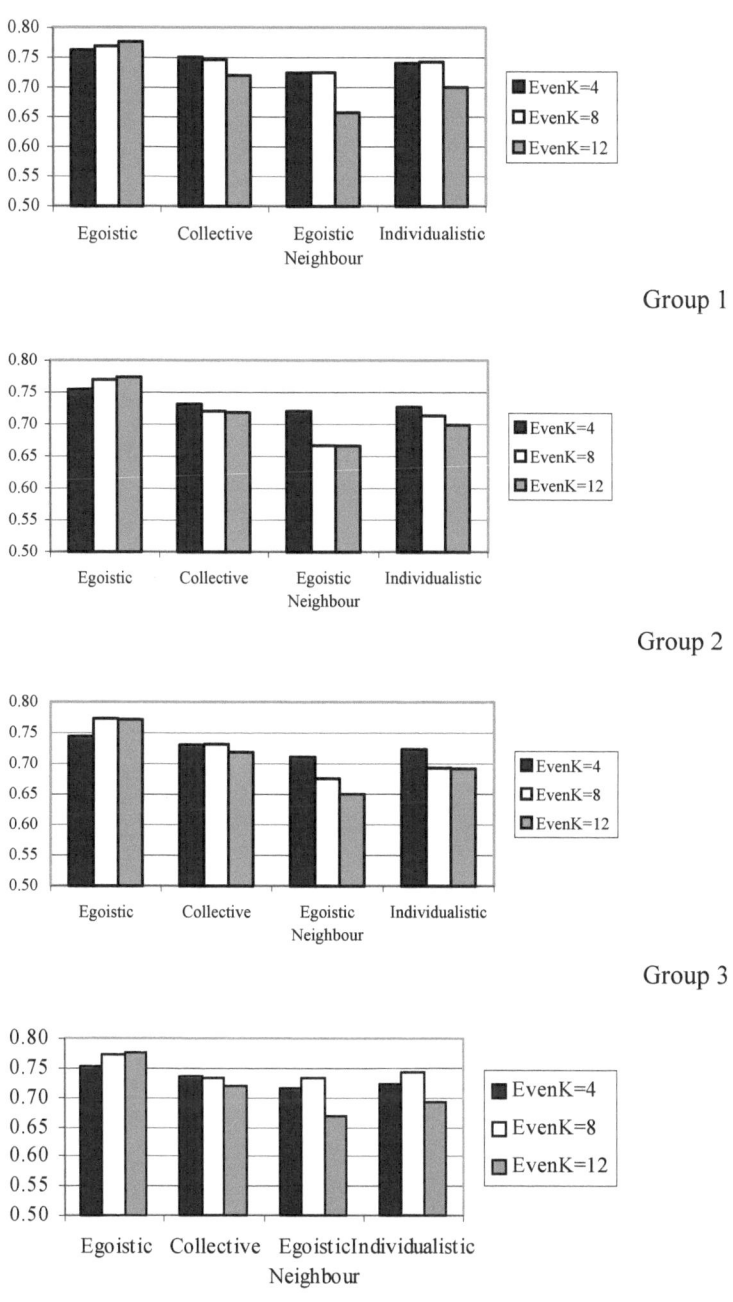

Fig. A6.5. Group payoff structures for selected parameters (bifurcation)

List of figures

Fig. 1.1. Spatial industry distributions - German pharmaceutical biotechnology ..8
Fig. 1.2. Continuity and change in cluster development 10
Fig. 3.1. Italian Industrial Districts – Main locations 48
Fig. 3.2. Porterian Clusters – the constituent elements 50
Fig. 4.1. The nature of ideal-typical clusters 61
Fig. 4.2. Cyclical interaction of micro-, meso- and macro-level in clusters 63
Fig. 4.3. Studying cluster development – Timeframe, assumptions and tools 74
Fig. 5.1. Possible system configurations for N=3, A_N=2...86
Fig. 5.2. System configurations and fitness levels for N=3 and K=0 87
Fig. 5.3. System configurations and fitness levels for N=3 and K=1 87
Fig. 5.4. System configurations and fitness levels for N=3 and K=2 88
Fig. 6.1. Decomposable versus non-decomposable N/K systems 98
Fig. 6.2 Near decomposable N/K systems...98
Fig. 6.3. Elements, functions and interdependence 99
Fig. 7.1. Agent, group and system dynamics in clusters 110
Fig. 7.2. Clusters as co-evolving systems – Applying the N/K framework.........112
Fig. 7.3. The model in schematic representation...115
Fig. 7.4. Division of labour and N/K(C) parameters 121
Fig. 7.5. Selected values of 'EvenK' and corresponding fitness landscapes.......123
Fig. 7.6. Search (- -) and selection (__) landscape in individualistic clusters 126
Fig. 7.7. Search (- -) and selection (__) landscape in collective clusters............127
Fig. 7.8. Search (- -) and selection (__) landscape in alliance clusters...............128
Fig. 7.9. Search (- -) and selection (__) landscape in leader firm clusters.........129
Fig. 8.1. Landscape complexity, bifurcation (steps 600 and 1200) and cluster
adaptation - Example runs for EvenK=4 and EvenK=14 138
Fig. 8.2. Division of labour, co-ordination and adaptive performance...............139
Fig. 8.3. Division of labour, co-ordination and stability in adaptation...............140
Fig. 8.4. Search and selection landscapes for leader firm clusters (EvenK=18) .143
Fig. 8.5. Agent numbers and adaptation: Individualistic and collective..............145
Fig. 8.6. Agent numbers and adaptation: Alliance and leader firm clusters........146
Fig. 8.7. Landscape complexity, perturbation (steps 600 and 1200) and cluster
adaptation - Example runs for EvenK=4 and 14 148
Fig. 8.8. Division of labour, co-ordination and adaptive performance...............149
Fig. 8.9. Division of labour, co-ordination and stability in adaptation...............150
Fig. 8.10. Agent numbers and adaptation: Individualistic and collective............151
Fig. 8.11. Agent numbers and adaptation: Alliance and leader firm clusters......152
Fig. 8.12. Group behaviour and fitness – Prisoner's dilemma revisited..............158

Fig. A6.1. Landscape complexity and adaptive performance (bifurcation) 206
Fig. A6.2. Landscape complexity and stability in adaptation (bifurcation) 206
Fig. A6.3. Landscape complexity and adaptive performance (perturbation) 213
Fig. A6.4. Landscape complexity and stability of adaptation (perturbation) 213
Fig. A6.5. Group payoff structures for selected parameters (bifurcation)..........221

List of tables

Table 8.1. Governance structure and average group fitness (bifurcation) 155
Table 8.2. Governance structure and average group fitness (perturbation) 156

Table A5.1. Model object structure 189
Table A5.2. Agent orientations (reflecting different modes of co-ordination) ...191
Table A5.3. Simulation settings..191
Table A5.4. Objects, parameters and variables..192
Table A5.5. Variables and equations 197
Table A6.1. Co-ordination mechanisms in the different populations 203
Table A6.2. Landscape complexity and average system fitness (bifurcation)....204
Table A6.3. Landscape complexity and standard deviation of system fitness
(bifurcation)..205
Table A6.4. Adaptation and group fitness (bifurcation) 207
Table A6.5. Landscape complexity and average system fitness
(perturbation)...211
Table A6.6. Landscape complexity and standard deviation of system fitness
(perturbation)...212
Table A6.7. Adaptation and group fitness (perturbation) 214
Table A6.8. Payoff for individualistic and egoistic groups (bifurcation) 217
Table A6.9. Payoff for individualistic and egoistic groups (perturbation) 219

References

Acemoglu D, Aghion P, Griffith R and Zilibotti F (2004) Vertical Integration and Technology: Theory and Evidence. Working Paper No. W10997. National Bureau of Economic Research (NBER), Cambridge, MA

Allen PM (1997) Cities and Regions as Self-Organizing Systems. Gordon and Breach Science Publishers, Amsterdam

Altenberg L (1994) Evolving Better Representations through Selective Genome Growth. In: Proceedings of the IEEE World Congress on Computational Intelligence, Pistcataway, NJ

Altenberg L (1995) Genome Growth and the Evolution of the Genotype-Phenotype Map. In: Banzhaf W and Eckman FH (eds) Evolution and Biocomputation. Springer, Berlin and Heidelberg, pp 205-259

Altenberg L (1997) Nk Fitness Landscapes. In: Back T, Fogel D and Michalewicz Z (eds) The Handbook of Evolutionary Computation. Oxford University Press, Oxford et al., pp B2.7:5—B2.7:10

Amin A (1994) Case Study III: Santa Croce in Context or How Industrial Districts Respond to the Restructuring of World Markets. In: Leonardi R and Nanetti RY (eds) Regional Development in a Modern European Economy: The Case of Tuscany. Pinter, London, New York, pp 170-186

Amin A (2000) Industrial Districts. In: Sheppard E and Barnes TJ (eds) A Companion to Economic Geography. Blackwell, Oxford et al., pp 149-168

Amin A and Cohendet P (2000) Organisational Learning and Governance through Embedded Practices. Journal of Management and Governance 4: 93-116

Amin A and Robins K (1990) Industrial Districts and Regional Development : Limits and Possibilities. In: Pyke F, Becattini G and Sengenberger W (eds) Industrial Districts and Inter-Firm Co-Operation in Italy. International Institute for Labour studies, Geneva, pp 185-220

Amin A and Thrift N (1992) Neo-Marshallian Nodes in Global Networks. International Journal of Urban and Regional Research 571-587

Anderson D (1992) A Linkage Approach to Industrial Location. Growth and Change 23: 321-334

Anselin L, Varga A and Acs Z (1997) Local Geographic Spillovers between University Research and High Technology Innovations. Journal of Urban Economics 42: 422-448

Arthur WB (1989a) Competing Technologies, Increasing Returns, and Lock-in by Historical Events. Economic Journal 99: 116-131

Arthur WB (1989b) Industry Location Patterns and the Importance of History. Working Paper No. 84. Centre for Economic Policy Research, Stanford University, Stanford, CA

Arthur WB (1990) Silicon Valley Locational Clusters: When Do Increasing Returns Imply Monopoly? Mathematic Social Sciences 19: 116-131

Asheim BT (1996) Industrial Districts as 'Learning Regions'. A Condition for Prosperity. European planning studies 4: 379-400

Audretsch DB and Cooke P (2001) Die Entwicklung regionaler Biotechnologie Cluster in den USA und Großbritannien (The Development of Regional Biotechnology Clusters in the United States and Britain). Working Paper No. 107. Akademie für Technologie-folgenabschätzung, Stuttgart

Audretsch DB and Feldman MP (1999) Innovation in Cities: Science-Based Diversity, Specialization and Localized Competition. European Economic Review 43: 409-429

Audretsch DB and Feldman MP (1996) R&D Spillovers and the Geography of Innovation and Production. American Economic Review 86: 630-640

Audretsch DB and Feldman MP (1995) Innovative Clusters and the Industry Life Cycle. Working Paper No. 1161. Centre for Economic Policy Research (CEPR), London

Audretsch DB and Fritsch M (1999) The Industry Component of Regional New Firm Formation Processes. Review of Industrial Organization 15: 239-252

Audretsch DB and Stephan PE (1996) Company-Scientist Locational Links: The Case of Biotechnology. American Economic Review 86: 641-652

Auerswald P, Kauffman S, Lobo J and Shell K (2000) The Production Recipes Approach to Modeling Technological Innovation: An Application to Learning by Doing. Journal of Economic Dynamics & Control 24: 389-450

Autant-Bernard C (2003) Specialisation, Diversity and Geographical Diffusion of Knowledge. In: Danish Research Unit on Industrial Dynamics (DRUID) Summer Conference Creating, Sharing and Transferring Knowledge, The role of Geography, Institutions and Organizations, Copenhagen, Denmark

Axelrod R (1997) The Complexity of Co-Operation. Princeton University Press, Princeton, NJ

Axelrod R and Cohen MD (1999) Harnessing Complexity: Organizational Implications of a Scientific Frontier. The Free Press, New York

Aydalot P (1986) The Location of New Firm Creation: The French Case. In: Keeble D and Wever E (eds) New Firms and Regional Development in Europe. Croom Helm, London, pp 105-123

Bagnasco A (1977) Tre Italie: La Problematica Territoriale dello Sviluppo Italiano. Il Mulino, Bologna

Baldwin CY, Clark KB and Woodard CJ (2003) The Pricing and Profitability of Modular Clusters. Electronic Working Paper August 2003, Harvard Business School

Baldwin RE (1999) Agglomeration and Endogenous Capital. European Economic Review 43: 253-280

Baptista R (1998) Clusters, Innovation, and Growth: A Survey of the Literature. In: Swann P and Prevezer M (eds) The Dynamics of Industrial Clustering: International Comparisons in Computing and Biotechnology. Oxford University Press, Oxford, pp 13-51

Baptista R (2000) Do Innovations Diffuse Faster within Geographical Clusters? International Journal of Industrial Organization 18: 515-535

Baranes E and Tropeano J-P (2003) Why Are Technological Spillovers Spatially Bounded? A Market Oriented Approach. Regional Science and Urban Economics 33: 445-466

Bathelt H, Malmberg A and Maskell P (2002) Cluster and Knowledge: Local Buzz, Global Pipelines and the Process of Knowledge Creation. Working Paper No. 02-12. Danish Research Unit for Industrial Dynamics (DRUID), Aalborg, Denmark

Beal BD and Gimeno J (2001) Geographic Agglomeration, Knowledge Spillovers, and Competitive Evolution. Working Paper No. 2001/26/SM. INSEAD, Fontainebleu France

Beaudry C and Breschi S (2000) Does 'Clustering' Really Help Firms' Innovative Activities? Working Paper No. 111. Centre of Research on Innovation and Internationalization (CESPRI), Milan

Becattini G (1990) The Marshallian Industrial District as a Socio-Economic Notion. In: Pyke F, Becattini G and Sengenberger W (eds) Industrial Districts and Inter-Firm Co-Operation in Italy. International Institute for Labour Studies, Geneva, pp 37-51

Becattini G (1994) The Development of Light Industry in Tuscany: An Interpretation. In: Leonardi R and Nanetti RY (eds) Regional Development in a Modern European Economy: The Case of Tuscany. Pinter Publishers, London, New York, pp 69-86

Becattini G (2002) From Marshall's to the Italian Industrial Districts. A Brief Critical Reconstruction. In: Curzio AQ and Fortis M (eds) Complexity and Industrial Clusters - Dynamics and Models in Theory and Practice. Physica, Heidelberg, pp 83-106

Bellandi M (1996) Research Briefing: Innovation and Change in the Marshallian Industrial Districts. European Planning Studies 4: 357-368

Belussi F (1999) Path-Dependency vs. Industrial Dynamics: An Analysis of Two Heterogeneous Districts. Human Systems Management 18: 161-174

Belussi F, Gottardi G and Rullani E (2003) The Technological Evolution of Industrial Districts. Kluwer Academic Publishers, Boston

Bhaskar R (1998) The Possibility of Naturalism. A Philosophical Critique of the Contemporary Human Sciences. Routledge, London

Bianchi G (1998) Requiem for the Third Italy? Rise and Fall of a Too Successful Concept. Entrepreneurship & Regional Development 10: 93-116

Bischi G-I, Dawid H and Kopel M (2003) Spillover Effects and the Evolution of Firm Clusters. Journal of Economic Behavior & Organization 50: 47-75

Bluestone B and Harrison B (1982) Deindustrialization of America. Plant Closings, Community Abandonment and the Dismantling of Basic Industry. Basic Books, New York

Boccura N (2004) Modelling Complex Systems. Springer, New York

Boero R, Castellani M and Squazzoni F (2004) Micro Behavioural Attitudes and Macro Technological Adaptation in Industrial Districts: An Agent-Based Prototype. Journal of Artificial Societies and Social Simulation 7: n.a.

Boje DM and Whetten DA (1981) Effects of Organizational Strategies and Contextual Constraints on Centrality and Attributions of Influence in Interorganizational Networks. Administrative Science Quarterly 26: 378-395

Boschma RA and Frenken K (2005) Why Is Economic Geography Not an Evolutionary Science? Towards an Evolutionary Economic Geography. Working Paper No. 05.01. Papers in Evolutionary Economic Geography, Utrecht University

Boschma RA and Lambooy JG (1999) Evolutionary Economics and Economic Geography. Journal of Evolutionary Economics 9: 411-429

Boschma RA and Lambooy JG (2002) Knowledge, Market Structure, and Economic Coordination: Dynamics of Industrial Districts. Growth and Change 33: 291-312

Bottazzi L and Peri G (2002) Innovation and Spillovers in Regions: Evidence from European Patent Data. Working Paper No. 215. Innocenzo Gasparini Institute for Economic Research (IGIER), Bocconi University, Milan

Braczyk H-J, Cooke P and Heidenreich M (1998) Regional Innovation Systems. University College London (UCL) Press, London

Braunerhjelm P, Carlsson B, Cetindamar D and Johansson D (2000) The Old and the New, the Evolution of Polymer and Biomedical Clusters in Ohio and Sweden. Journal of Evolutionary Economics 10: 471-488

Brenner T (2000) The Evolution of Localised Industrial Clusters: Identifying the Processes of Self-Organisation. Working Paper No. 0011. Max Planck Institute for Research into Economic Systems, Jena

Brenner T (2001) Self-Organisation, Local Symbiosis and the Emergence of Localised Industrial Clusters. Working Paper No. 0103. Max Planck Institute for Research into Economic Systems, Jena

Brenner T (2003) Policy Measures to Support the Emergence of Localised Industrial Clusters. In: Fornahl D and Brenner T (ed) Cooperation, Networks, and Institutions in Regional Innovation Systems. Edward Elgar, Cheltenham, pp 325-350

Brenner T (2004) Local Industrial Clusters, Existence, Emergence and Evolution. Routledge, London

Brenner T and Weigelt N (2001) The Evolution of Industrial Clusters - Simulating Spatial Dynamics. Advances in Complex Systems 4: 127-147

Breschi S (1998) Agglomeration Economies, Knowledge Spillovers, Technological Diversity and Spatial Clustering of Innovations. Working Paper No. 57. LIUC Papers in Economics, Cattaneo University, Castellanza Italy

Breschi S and Lissoni F (2001) Knowledge Spillovers and Local Innovation Systems: A Critical Survey. Working Paper No. 84. LIUC Papers in Economics, Cattaneo University, Castellanza Italy

Bresnahan TF, Gambardella A and Saxenian A (2001) 'Old Economy' Inputs for 'New Economy' Outcomes: Cluster Formation in the New Silicon Valleys. Industrial and Corporate Change 10: 835-860

Bresnahan TF and Malerba F (1999) Industrial Dynamics and the Evolution of Firms' and Nations' Competitive Capabilities in the World Computer Industry. In: Mowery DC and Nelson RR (eds) Sources of Industrial Leadership. Cambridge University Press, Cambridge, Mass., pp 79-132

Brülhart M (1996) Commerce Et Spécialization Géographique Dans L'Union Européenne. Economie Internationale 65: 169-202

Brusco S (1982) The Emilian Model: Productive Decentralisation and Social Integration. Cambridge Journal of Economics 6: 167-184

Brusco S (1986) Small Firms and Industrial Districts: The Experience of Italy. In: Keeble D and Wever E (eds) New Firms and Regional Development in Europe. Croom Helm, London, pp 184-202

Brusco S (1990) The Idea of the Industrial District: It's Genesis. In: Pyke F, Beccattini G and Sengenberger W (eds) Industrial Districts and Inter-Firm Co-Operation in Italy. International Institute for Labour Studies, Geneva, pp 10-19

Brusco S and Righi E (1989) Local Government, Industrial Policy and Social Consensus: The Case of Modena (Italy). Economy and society 18: 405-424

Bull AC, Pitt M and Szarka J (1991) Small Firms and Industrial Districts, Structural Explanations of Small Firm Viability in Three Countries. Entrepreneurship & Regional Development 3: 83-99

Burt RS (2001) Bandwidth and Echo: Trust, Information and Gossip in Social Networks. In: Rauch JE and Casella A (eds) Networks and Markets. Russell Sage Foundation, New York, pp 30-74

Burt RS (1992) Structural Holes - the Social Structure of Competition. Harvard University Press, Cambridge MA

Byrne DS (1998) Complexity Theory and the Social Sciences. Routledge, London

Caeldries F (1996) The Institutional Embeddedness of Strategy: Predation through Legislation (or, See You in Court). Advances in Strategic Management 13: 215-24

Cainelli G and Zoboli R (2004a) The Structural Evolution of Industrial Districts and Adaptive Competitive Advantages. In: Cainelli G and Zoboli R (eds) The Evolution of Industrial Districts: Changing Governance, Innovation and Internationalisation of Local Capitalism in Italy. Physica, Heidelberg, New York, pp 1-29

Cainelli G and Zoboli R (2004b) The Evolution of Industrial Districts: Changing Governance, Innovation and Internationalisation of Local Capitalism in Italy. Physica, Heidelberg

Camagni R (1991) Innovation Networks. Belhaven Press, London

Capecchi V (1990) A History of Flexible Specialisation and Industrial Districts in Emilia-Romagna. In: Pyke F, Beccattini G and Sengenberger W (eds) Industrial Districts and Inter-Firm Co-Operation in Italy. International Institute for Labour Studies, Geneva, pp 20-36

Cappellin R (2003) Networks and Technological Change in Regional Clusters. In: Bröcker J, Dohse D and Soltwedel R (eds) Innovation Clusters and Interregional Competition. Springer, Berlin and Heidelberg, pp 52-78

Carley KM and Lee J-S (1998) Dynamic Organizations: Organizational Adaptation in a Changing Environment. Advances in Strategic Management 15: 269-297

Carley KM and Svoboda DM (1996) Modeling Organizational Adaptation as a Simulated Annealing Process. Sociological Methods and Research 25: 138-168

Carney J (1980) Regions in Crisis - New Perspectives in European Regional Theory. Croom Helm, London

Casper S (2000) Institutional Adaptiveness, Technology Policy, and the Diffusion of New Business Models: The Case of German Biotechnology. Organization Studies 21: 887-914

Chamberlin EH (1933) The Theory of Monopolistic Competition. Harvard University Press, Cambridge MA

Chang M-H and Harrington JE (2000) Centralization vs. Decentralization in a Multi-Unit Organization: A Computational Model of a Retail Chain as a Multi-Agent Adaptive System. Management Science 46: 1427-1440

Chang M-H and Harrington JE (2004) Organization of Innovation in a Multi-Unit Firm: Coordinating Adaptive Search on Multiple Rugged Landscapes. In: Barnett WA, Deissenberg C and Feichtinger G (eds) Economic Complexity: Non-Linear Dynamics, Multi-Agents Economies, and Learning. Elsevier, Amsterdam, pp 189-214

Christaller W (1966) Central Places in Southern Germany. Prentice-Hall, Englewood Cliffs, N.J.

Coase RH (1937) The Nature of the Firm. Economica 4: 386-405

Combes PP (1997) Industrial Agglomeration under Cournot Competition. Annales d'Économie et de Statistique 45: 161-182

Cooke P (1997) Regions in a Global Market: The Experiences of Wales and Baden-Württemberg. Review of International Political Economy 4: 349-381

Cooke P, Gomez Uranga M and Extebarria G (1997) Regional Innovation Systems: Institutional and Organisational Dimensions. Research Policy 26: 475-491

Coriat B and Dosi G (2002) The Institutional Embeddedness of Economic Change: An Appraisal of the 'Evolutionary' and the 'Regulationist' Research Programmes. In: Hogdson GM (ed) A Modern Reader in Institutional and Evolutionary Economicy: Key Concepts. Edward Elgar, Cheltenham, pp 95-123

Courlet C and Soulage B (1995) Industrial Dynamics and Territorial Space. Entrepreneurship & Regional Development 7: 287-307

Courtney L (1878) The Migration of Centres of Industrial Energy. The Fortnightly Review n.a.

Cowan R and Jonard N (2003) The Dynamics of Collective Invention. Journal of Economic Behavior & Organization 52: 513-532

Crafts N and Mulatu A (2004) What Explains the Location of Industry in Britain, 1871-1931? Working Paper No. 4356. Centre for Economic Policy Research (CEPR), London

Curzio AQ and Fortis M (2002) Complexity and Industrial Clusters - Dynamics and Models in Theory and Practice. Physica, Heidelberg

Dahl MS and Pedersen CØR (2003) Knowledge Flows through Informal Contacts in Industrial Clusters: Myths or Realities? Working Paper No. 03-01. Danish Research Unit for Industrial Dynamics (DRUID), Aalborg, Denmark

Dalum B, Pedersen CØR and Villumsen G (2002) Technological Life Cycles: Regional Clusters Facing Disruption. Working Paper No. 02-10. Danish Research Unit for Industrial Dynamics (DRUID), Aalborg, Denmark

Darwin C (1859) On the Origin of Species by Means of Natural Selection, Or: The Preservation of Favoured Races in the Struggle for Life. J. Murray, London

Davelaar E-J and Nijkamp P (1990) Technological Innovation and Spatial Transformation. Technological Forecasting and Social Change 37: 181-202

David PA (1985) Clio and the Economic of QWERTY. American Economic Review Papers and Proceedings 75: 332-337

Davis DR and Weinstein DE (2002) Bones, Bombs, and Break Points: The Geography of Economic Activity. American Economic Review 92: 1269-1289

Davis DR and Weinstein DE (2003) Market Access, Economic Geography and Comparative Advantage: An Empirical Test. Journal of International Economics 59: 1-23

Dei Ottati G (1994a) Case Study I: Prato and Its Evolution in a European Context. In: Leonardi R and Nanetti RY (eds) Regional Development in a Modern European Economy: The Case of Tuscany. Pinter Publishers, London, pp 116-145

Dei Ottati G (1994b) Cooperation and Competition in the Industrial District as an Organization Model. European Planning Studies 2: 463-483

Dei Ottati G (1994c) Trust, Interlinking Transactions and Credit in the Industrial District. Cambridge Journal of Economics 18: 529-546

Dei Ottati G (1996) The Remarkable Resilience of the Industrial Districts of Tuscany. Working Paper No 28. ERSC Centre for Business Research, University of Cambridge, Cambridge UK

Dixit A and Stiglitz J (1977) Monopolistic Competition and Optimum Product Diversity. American Economic Review 67: 297-308

Dosi G, Freeman C, Nelson RR, Silverberg G and Soete L (1988) Technical Change and Economic Theory. Pinter Publishers, London

Dosi G, Levinthal DA and Marengo L (2003) Bridging Contested Terrain: Linking Incentive-Based and Learning Perspectives on Organizational Evolution. Industrial and Corporate Change 12: 413-435

Dosi G and Marengo L (2003) Division of Labor, Organizational Coordination and Market Mechanism in Collective Problem-Solving. Working Paper No. 2003/04. Laboratory of Economics and Management (LEM), Sant'Anna School of Advanced Studies, Pisa

Dumais G, Ellison G and Glaeser EL (2002) Geographic Concentration as a Dynamic Process. The Review of Economics and Statistics 84: 193-204

Edquist C (1997) Systems of Innovation. Pinter Publishers, London

Ellison G and Glaeser EL (1994) Geographic Concentration in U.S. Manufacturing Industries: A Dartboard Approach. Working Paper No. 4840. National Bureau of Economic Research (NBER), Cambridge, MA

Ellison G and Glaeser EL (1997) Geographic Concentration in U.S. Manufacturing Industries: A Dartboard Approach. Journal of Political Economy 105: 889-927

Ellison G and Glaeser EL (1999) The Geographic Concentration of Industry: Does Natural Advantage Explain Agglomeration? American Economic Review 89: 311-327

Engländer O (1926) Kritisches und Positives zu einer allgemeinen, reinen Lehre vom Standort (A Critical and Positive Account of Pure Location Choice Theory). Zeitschrift für Volkswirtschaft und Sozialpolitik 5: 435 - 505

Engländer O (1924) Theorie des Güterverkehrs und der Frachtsätze (A Theory of Transportation and Transportation Cost). G. Fischer, Jena

Enright MJ (1995) Organization and Coordination in Geographically Concentrated Industries. In: Lamoreaux NR and Raff DMG (eds) Coordination and Information - Historical Perspecives on the Organization of Enterprise. The University of Chicago Press, Chicago, pp 103-146

Enright MJ (2003) Regional Clusters: What We Know and What We Should Know. In: Bröcker J, Dohse D and Soltwedel R (eds) Innovation Clusters and Interregional Competition. Springer, Heidelberg, pp 99-129

Ethiraj SK and Levinthal DA (2004a) Bounded Rationality and the Search for Organizational Architecture: An Evolutionary Perspective on the Design of Organizations and Their Evolvability. Administrative Science Quarterly 49: 404-437

Ethiraj SK and Levinthal DA (2004b) Modularity and Innovation in Complex Systems. Management Science 50: 159-173

Fehr E and Fischbacher U (2002) Why Social Preferences Matter: The Impact of Non-Selfish Motives on Competition, Cooperation and Incentives. Economic Journal 112: C1-C33

Feldman MP (1999) The New Economics of Innovation, Spillovers and Agglomeration: A Review of Empirical Studies. Economics of Innovation and New Technology 8: 5-25

Feldman MP and Schreuder Y (1996) Initial Advantage: The Origins of the Geographic Concentration of the Pharmaceutical Industry in the Mid-Atlantic Region. Industrial and Corporate Change 5: 839-862

Fleming L and Sorenson O (2001) Technology as a Complex Adaptive System: Evidence from Patent Data. Research Policy 30: 1019-1039

Flier B, Van Den Bosch FAJ and Volberda HW (2003) Co-Evolution in Strategic Renewal Behaviour of British, Dutch and French Financial Incumbents: Interaction of Environmental Selection, Institutional Effects and Managerial Intentionality. Journal of Management Studies 40: 2163-2187

Fornahl D and Brenner T (2003) Cooperation, Networks, and Institutions in Regional Innovation Systems. Edward Elgar, Cheltenham

Frenken K (2001) Understanding Product Innovation Using Complex Systems Theory. Ph.D. Dissertation, University of Amsterdam and University of Grenoble

Frenken K (2005) Innovation, Evolution and Complexity Theory. Edward Elgar, Cheltenham

Frenken K (n.a.) Technological Innovation and Complexity Theory. Economics of Innovation and New Technology (forthcoming)

Frenken K, Marengo L and Valente M (1999) Interdependencies, near-Decomposability and Adaptation. In: Brenner T (ed) Computational Techniques for Modelling Learning in Economics. Kluwer Academic Publishers, Boston, pp 145-165

Frenken K and Nuvolari A (2003) The Early Development of the Steam Engine: An Evolutionary Interpretation Using Complexity Theory. Working Paper No. 03.15. Eindhoven Centre for Innovation Studies, Eindhoven NL

Frenken K and Nuvolari A (2004) The Early History of Steam Engine Technology: An Evolutionary Interpretation Using Complexity Theory. Industrial and Corporate Change 13: 419-450

Frenken K and Valente M (2003) Complexity, Patches and Fitness: A Model of Decentralised Problem-Solving. In: 8th Workshop on Economics of Heterogeneous Interacting Agents (WEHIA), University of Kiel

Fritsch M and Franke G (2004) Innovation, Regional Knowledge Spillovers and R&D Co-Operation. Research Policy 33: 245-255

Fujita M, Krugman P and Venables AJ (1999) The Spatial Economy: Cities, Regions, and International Trade. MIT Press, Cambridge, MA

Fujita M and Thisse J-F (1996) Economics of Agglomeration. Journal of the Japanese and International Economies 10: 339-378

Fujita M and Thisse J-F (2002) Economies of Agglomeration. Cities, Industrial Location and Regional Growth. Cambridge University Press, Cambridge UK

Gabe T (2003) Local Industry Agglomeration and New Business Activity. Growth and Change 34: 17-39

Galaskiewicz J and Wasserman S (1981) A Dynamic Study of Change in a Regional Coporate Network. American Sociological Review 46: 475-484

Garnsey E (1998) The Genesis of the High Technology Milieu: A Study in Complexity. International Journal of Urban and Regional Research 361-377

Gavetti G and Levinthal D (2000) Looking Forward and Looking Backward: Cognitive and Experiential Search. Administrative Science Quarterly 45: 113-137

Gell-Mann M (2002) What Is Complexity? In: Curzio AQ and Fortis M (eds) Complexity and Industrial Clusters: Dynamics and Models in Theory and Practice. Physica, Heidelberg, pp 13-24

Gertler MS (1996) Worlds Apart: The Changing Market Geography of the German Machinery Industry. Small Business Economics 8: 87-106

Glaeser EL (2003) Reinventing Boston: 1640-2003. Working Paper No. 10166. National Bureau of Economic Research (NBER), Cambridge, MA

Glaeser EL, Dumais G and Ellison G (2002) Geographic Concentration as a Dynamic Process. Review of Economics and Statistics 84: 193-204

Glaser EL, Laibson D, Scheinkman JA and Soutter C (1999) What Is Social Capital? The Determinants of Trust and Trustworthiness. Working Paper No. 7216. National Bureau of Economic Research, Cambridge, MA

Goldstein GS and Gronberg TJ (1984) Economies of Scope and Economies of Agglomeration. Journal of Urban Economics 16: 91-104

Goodman EJ, Bamford J and Saynor P (1989) Small Firms and Industrial Districts in Italy. Routledge, London

Grabher G (1993) The Weakness of Strong Ties: The Lock-in of Regional Development in the Ruhr Area. In: Grabher G (ed) The Embedded Firm - on the Socioeconomics of Industrial Networks. Routledge, London, New York, pp 255-277

Granovetter M (1985) Economic Action and Social Structure: The Problem of Embeddedness. American Journal of Sociology 91: 481-510

Granovetter M (1973) The Strength of Weak Ties. American Journal of Sociology 78: 1360-1380

Greenwood R and Hinings CR (1996) Understanding Radical Organizational Change: Bringing Together the Old and the New Institutionalism. Academy of Management Review 21: 1022-1054

Grossman GM and Helpman E (1991) Innovation and Growth in the Global Economy. MIT Press, Cambridge, MA

Guerrieri P and Pietrobelli C (2000) Models of Industrial Districts' Evolution and Changes in Technological Régimes. In: Danish Research Unit for Industrial Dynamics (DRUID): The Learning Economy - Firms, Regions and Nation Specific Institutions, Rebild, Denmark

Guerrieri P and Pietrobelli C (2001) Models of Industrial Clusters' Evolution and Changes in Technological Régimes. In: Guerrieri P, Iammarino S and Pietrobelli C (eds) The Global Challenge to Industrial Districts: Small and Medium Sized Enterprises in Italy and Taiwan. Edward Elgar, Cheltenham, pp 11-34

Guerrieri P, Iammarino S and Pietrobelli C (2001) The Global Challenge to Industrial Districts: Small and Medium Sized Enterprises in Italy and Taiwan. Edward Elgar, Cheltenham

Hanson GH (1996a) Agglomeration, Dispersion, and the Pioneer Firm. Journal of Urban Economics 39: 255-281

Hanson GH (1996b) Economic Integration, Intraindustry Trade, and Frontier Regions. European Economic Review 40: 941-949

Harrigan J and Venables AJ (2004) Timeliness, Trade and Agglomeration. Working Paper No. 4294. Centre for Economic Policy Research (CEPR), London

Harrisson B (1992) Industrial Districts: Old Wine in New Bottles? Regional Studies 26: 469-483

Hayter R (1997) The Dynamics of Industrial Location. The Factory, the Firm and the Production System. Wiley, New York

Head K and Mayer T (2003) The Empirics of Agglomeration and Trade. Working Paper No. 3985. Centre for Economic Policy Research (CEPR), London

Head K, Ries J and Swenson D (1995) Agglomeration Benefits and Location Choice: Evidence from Japanese Manufacturing Investments in the United States. Journal of International Economics 38: 223-247

Heiduk G and Pohl N (2002) Silicon Valley's Innovative Milieu: A Cultural Mix of Entrprepreneurs/ an Entrepreneurial Mix of Cultures. Erkunde 56: 241-252

Helpman E (1997) The Size of Regions. In: Pines D, Sadka E and Zilcha I (eds) Topics in Public Economics. Theoretical and Applied Analysis. Cambridge University Press, Cambridge UK, pp 33-54

Helpman E and Krugman P (1985) Market Structure and Foreign Trade. MIT Press, Cambridge, MA

Helsley RW and Strange WC (2002) Innovation and Input Sharing. Journal of Urban Economics 51: 25-45

Helsley RW and Strange WC (2003) Agglomeration, Opportunism, and the Organization of Production. Working Paper No. 03-02. Centre for Urban Economics and Real Estate (CUER), Vancouver

Henderson JV (2003) Marshall's Scale Economies. Journal of Urban Economics 53: 1-28

Hirschmann AO (1967) Die Strategie der wirtschaftlichen Entwicklung (The Strategy of Economic Development). Fischer Publishing, Stuttgart

Holland JH (2002) Complex Adaptive Systems and Spontaneous Emergence. In: Curzio AQ and Fortis M (eds) Complexity and Industrial Clusters: Dynamics and Models in Theory and Practice. Physica, Heidelberg and New York, pp 25-34

Holländer H (1990) A Social Exchange Approach to Voluntary Cooperation. American Economic Review 80: 1157-1167

Holmes TJ (1999) How Industries Migrate When Agglomeration Economies Are Important. Journal of Urban Economics 45: 240-263

Hoover EM (1937) Location Theory and the Shoe and Leather Industries. Harvard University Press, Cambridge MA

Hoover EM (1970) An Introduction to Regional Economics. A. A. Knopf, New York

Hovhannisian K and Valente M (2004) Modeling Directed Local Search Strategies on Technology Landscapes: Depth and Breadth. Working Paper No. 91. University of Trento, Research on Organizations, Coordination and Knowledge (ROCK), Trento Italy

Isaksen A (2003) 'Lock-in' of Regional Clusters: The Case of Offshore Engineering in the Oslo Region. In: Fornahl D and Brenner T (eds) Cooperation, Networks and Institutions in Regional Innovation Systems. Edward Elgar, Cheltenham, pp 247-276

Jacobs J (1969) The Economy of Cities. Vintage, New York

Jaffe AB, Trajtenberg M and Henderson R (1993) Geographic Localisation of Knowledge Spillovers as Evidenced by Patent Citations. Quarterly Journal of Economics 10: 577-598

Jones RW and Kierzkowski H (2004) International Trade and Agglomeration: An Alternative Framework. Journal of Economics (Zeitschrift für Nationalökonomie) Suppl. 10: 163-177

Jovanovic MN (2003) Spatial Location of Firms and Industries: An Overview of Theory. Economia Internazionale 56: 23-82

Kandel E and Lazear EP (1992) Peer Pressure and Partnership. Journal of Political Economy 100: 801-817

Karlsson C and Johansson B (2005) Industrial Clusters and Inter-Firm Networks. Edward Elgar, Cheltenham

Kauffman SA (1993) The Origins of Order: Self-Organization and Selection in Evolution. Oxford University Press, Oxford

Kauffman SA, Lobo J and Macready WG (2000) Optimal Search on a Technology Landscape. Journal of Economic Behavior and Organization 43: 141-166

Kauffman SA and Macready WG (1995) Technological Evolution and Adaptive Organizations. Complexity 1: 26-43

Keller W (2000) Geographic Localization of International Technology Diffusion. Working Paper No. 7509. National Bureau of Economic Research (NBER), Cambridge, MA

Kern H (1996) Vertrauensverlust und blindes Vertrauen: Integrationsprobleme im ökonomischen Handeln (Loss of Trust and Blind Trust - Issues of Integration in Economic Action). SOFI-Mitteilungen 24: 7-14

Klepper S (1996) Entry, Exit, Growth, and Innovation over the Product Life Cycle. American Economic Review 86: 562-583

Klepper S (1997) Industry Life Cycles. Industrial and Corporate Change 6: 145-181

Klepper S (2002) The Capabilities of New Firms and the Evolution of the US Automobile Industry. Industrial and Corporate Change 11: 645-666

Klepper S and Sleeper S (2004) Entry by Spinoffs. Management Science (forthcoming)

Klimenko MM (2004) Competition, Matching, and Geographical Clustering at Early Stages of the Industry Life Cycle. Journal of Economics and Business 56: 177–195

Kline SJ and Rosenberg N (1986) An Overview of Innovation. In: Landau R and Rosenberg N (eds) The Positive Sum Strategy: Harnessing Technology for Economic Growth. National Academy Press, Washington DC, pp 275-306

Kollman K, Miller JH and Page SE (2000) Decentralization and the Search for Policy Solutions. Journal of Law, Economics, and Organization 16: 102-128

Kondra AZ and Hinings CR (1998) Organizational Diversity and Change in Institutional Theory. Organization Studies 19: 743-767

Koschatzky K (1998) Firm, Innovation and Region: The Role of Space in Innovation Processes. International Journal of Innovation Management 2: 383-408

Krugman P (1991a) Geography and Trade. MIT Press, Cambridge, MA

Krugman P (1991b) Increasing Returns and Economic Geography. Journal of Political Economy 99: 484-499

Krugman P and Venables AJ (1995) Globalization and the Inequality of Nations. Quarterly Journal of Economics 49: 857-880

Krugman P and Venables AJ (1996) Integration, Specialization and Adjustment. European Economic Review 40: 959-967

Kubon-Gilke G (1997) Verhaltensbindung und die Evolution ökonomischer Institutionen (Behaviour Constraints and the Evolution of Economic Institutions). metropolis, Marburg Germany

Lagendijk A and Cornford J (2000) Regional Institutions and Knowledge - Tracking New Forms of Regional Development Policy. Geoforum 31: 209-218

Lane DA (2002) Complexity and Local Interactions: Towards a Theory of Industrial Districts. In: Curzio AQ and Fortis M (eds) Complexity and Industrial Clusters - Dynamics and Models in Theory and Practice. Physica, Heidelberg, pp 65-82

Langlois RN (2002) Modularity in Technology and Organization. Journal of Economic Behavior & Organization 49: 19-37

Lazerson MH (1990) Subcontracting in the Modena Knitwear Industry. In: Pyke F, Becattini G and Sengenberger W (eds) Industrial Districts and Inter-Firm Co-Operation in Italy. International Institute for Labour Studies, Geneva, pp 108-134

Lazonick W (1993) Industry Clusters versus Global Webs: Organizational Capabilities in the American Economy. Industrial and Corporate Change 2: 1-24

Le Bas C and Miribel F (2005) The Agglomeration Economies Associated with Information Technology Activities: An Empirical Study of the US Economy. Industrial and Corporate Change 14: 343-363

Lee C-M, Miller WF, Hancock MG and Rowen HS (2000) The Silicon Valley Edge - A Habitat for Innovation and Entrepreneurship. Stanford University Press, Stanford, CA

Lemarie S, Mangematin V and Torre A (2001) Is the Creation and Development of Biotech SMEs Localised? Conclusions Drawn from the French Case. Small Business Economics 17: 61-76

Levinthal DA (1997) Adaptation on Rugged Landscapes. Management Science 43: 934-950

Levinthal DA (2000) Organizational Capabilities in Complex Worlds. In: Dosi G, Nelson RR and Winter SG (eds) The Nature and Dynamics of Organizational Capabilities. Oxford University Press, Oxford, pp 363-379

Levinthal DA and March JG (1993) The Myopia of Learning. Strategic Management Journal 14: 95-112

Levinthal DA and Warglien M (1999) Landscape Design: Designing for Local Action in Complex Worlds. Organization Science 10: 342-357

List F (1842) Das nationale System der politischen Oekonomie. Erster Band. Der Internationale Handel, die Handelspolitik, und der Deutsche Zollverein (The National System of Political Economy. First Volume. International Trade, Trade Policy and Germany's Customs Agreement). Cotta'scher Verlag, Stuttgart

List F (1909) The National System of Political Economy. Longmans, Green and Co., London

Loasby BJ (2000) Organisations as Interpretative Systems. In: The Danish Research Unit of Industrial Dynamics (DRUID) Summer Conference: The Learning Economy - Firms, Regions and Nation Specific Institutions, Rebild, Denmark

Lombardi M (2003) The Evolution of Local Production Systems: The Emergence of the 'Invisible Mind' and the Evolutionary Pressures Towards More Visible 'Minds'. Research Policy 32: 1443-1462

López-Bazo E, Vayá E and Artís M (2004) Regional Externalities and Growth: Evidence from European Regions. Journal of Regional Science 44: 43-73

Lorenzen M and Foss NJ (2003) Cognitive Co-Ordination, Institutions and Clusters: An Exploratory Discussion. In: Fornahl D and Brenner T (eds) Cooperation, Networks and Institutions in Regional Innovation Systems. Edward Elgar, Cheltenham, pp 82-104

Lorenzen M and Maskell P (2005) The Cluster as a Nexus of Knowledge Creation. In: Cooke P and Piccaluga A (eds) Regional Economies as Knowledge Laboratories. Edward Elgar, Cheltenham, pp 77-92

Lösch A (1962) Die räumliche Ordnung der Wirtschaft (The Spatial Organisation of the Economy). Gustav Fischer Verlag, Stuttgart

Louch H (2000) Personal Network Integration: Transitivity and Homophily in Strong-Tie Relations. Social Networks 22: 45-64

Lundvall B-A (2002) The Learning Economy: Challenges to Economic Theory and Policy. In: Hogdson GM (ed) A Modern Reader in Institutional and Evolutionary Economics, Key Concepts. Edward Elgar, Cheltenham, pp 26-47

Lundvall B-A (1995) National Systems of Innovation: Towards a Theory of Innovation and Interactive Learning. Pinter, London

Lux T, Reitz S and Samanidou E (2005) Nonlinear Dynamics and Heterogeneous Interacting Agents. Springer, Berlin and Heidelberg

Maggioni MA (2002) Clustering Dynamics and the Location of High-Tech-Firms. Physica, Heidelberg

Maggioni MA (2004) Modeling the Structure and Evolution of Industrial Districts. In: Cainelli G and Zoboli R (eds) The Evolution of Industrial Districts: Changing Governance, Innovation and Internationalisation of Local Capitalism in Italy. Physica, Heidelberg, pp 78-113

Maillat D (1998a) From an Industrial District to the Innovative Milieu: Contribution to an Analysis of Territorised Productive Organisations. Recherches Économiques de Louvain 64: 111-129

Maillat D (1998b) Innovative Milieux and New Generations of Regional Policies. Entrepreneurship & Regional Development 10: 1-16

Malerba F (1999) Sectoral Systems of Innovation and Production. In: Danish Research Unit on Industrial Dynamics (DRUID) Summer Conference on: National Innovation Systems, Industrial Dynamics and Innovation Policy, Rebild, Denmark

Malmberg A and Maskell P (1997) Towards and Explanation of Regional Specialization and Industry Agglomeration. European Planning Studies 5: 25-41

Malmberg A and Maskell P (2001) The Elusive Concept of Localization Economies - Towards a Knowledge-Based Theory of Spatial Clustering. In: The Association of American Geographers' Annual Conference, New York

Malmberg A and Maskell P (2002) The Elusive Concept of Localization Economies: Towards a Knowledge-Based Theory of Spatial Clustering. Environment and Planning 34: 429-229

Marengo L and Dosi G (2005) Decentralization and Market Mechanisms in Collective Problem-Solving. Journal of Economic Behavior & Organization (forthcoming)

Marengo L, Dosi G, Legrenzi P and Pasquali C (2000) The Structure of Problem-Solving and the Structure of Organisations. Industrial and Corporate Change 9: 757-788

Markusen A (1996) Sticky Places in Slippery Space: A Typology of Industrial Districts. Economic Geography 72: 293-313

Marshall A (1920) Principles of Economics. Macmillan, London

Martin R (1999) The New 'Geographical Turn' in Economics: Some Critical Reflections. Cambridge Journal of Economics 23: 65-91

Martin R and Sunley P (1996) Paul Krugman's Geographical Economics and Its Implications for Regional Development Theory: A Critical Assessment. Economic Geography June 1996: 259-292

Martin R and Sunley P (2003) Deconstructing Clusters: Chaotic Concept or Policy Pancea? Journal of Economic Geography 3: 5-35

Maskell P (1999) The Firm in Economic Geography. Economic Geography 77: 1-13

Maskell P (2001) Towards a Knowledge-Based Theory of the Geographical Cluster. Industrial and Corporate Change 10: 921-943

Maskell P and Kebir L (2004) The Theory of the Cluster - What It Takes and What It Implies. In: Asheim BT, Cooke P and Martin R (eds) Clusters and Regional Development. Routledge, London (forthcoming)

Maskell P and Lorenzen M (2003) The Cluster as Market Organization. Working Paper No. 03-14. Danish Research Unit of Industrial Dynamics (DRUID), Aalborg, Denmark

Massey D and Meegan R (1982) Anatomy of Job Loss. The How, Why and Where of Employment Decline. Methuen, London

Mathias P (1983) The First Industrial Nation: An Economic History of Britain 1700-1914. Methuen, London

Maunier R (1908) La Distribution Géographique des Industries (The Geographical Distribution of Industries). Revue Internationale de Sociologie 16: 481-514

Maunier R (1909) La Localisation des Industries Urbaines (The Localisation of Urban Industries). Giard & Brière, Paris

McCann P and Sheppard S (2003) The Rise, Fall and Rise Again of Industrial Location Theory. Regional Studies 37: 649-663

McGahan AM (2004) How Industries Evolve: Principles for Achieving and Sustaining Superior Performance. Harvard Business School Press, Boston, MA

McKelvey M (1991) How Do National Systems of Innovation Differ?: A Critical Analysis of Porter, Freeman, Lundvall and Nelson. In: Hogdson GM and Screpanti E (eds) Rethinking Economics - Markets, Technology and Economic Evolution. Edward Elgar, Cheltenham, pp 117-137

McKelvey M, Alm H and Riccaboni M (2002) Does Co-Location Matter for Formal Knowledge Collaboration in the Swedish Biotechnology-Pharmaceutical Sector? Research Policy 1394: 1-19

Merry U (1999) Organizational Strategy on Different Landscapes: A New Science Approach. Systemic Practice and Action Research 12: 257-278

Meyer-Stamer J (1998) Path Dependence in Regional Development: Persistence and Change in Three Industrial Clusters in Santa Catarina, Brazil. World Development 26: 1495-1511

Morrison PCJ and Siegel DS (1999) Scale Economies and Industry Agglomeration Externalities: A Dynamic Cost Function Approach. American Economic Review 89: 272-290

Mothe JDL and Paquet G (1997) Local and Regional Systems of Innovation. Kluwer Academic Publishers, Boston

Moulaert F and Sekia F (2003) Territorial Innovation Models: A Critical Survey. Regional Studies 37: 289-302

Müller B, Finka M and Lintz G (2005) Rise and Decline of Industry in Central and Eastern Europe - a Comparative Study of Cities and Regions in Eleven Countries. Springer, Heidelberg

Murray JD (1993) Mathematical Biology. Springer, Heidelberg

Myrdal G (1957) Economic Theory and Under-Developed Regions. Ducksworth, London

Neary JP (2001) Of Hype and Hyperbolas: Introducing the New Economic Geography. Journal of Economic Literature XXXIX: 536-561

Nelson RR and Rosenberg N (1993) Technical Innovation and National Systems. In: Nelson RR (ed) National Innovation Systems - a Comparative Analysis. Oxford University Press, Oxford, pp 3-28

Nelson RR and Winter SG (1977) In Search of Useful Theory of Innovation. Research Policy 6: 36-76

Nelson RR and Winter SG (1982) An Evolutionary Theory of Economic Change. Harvard University Press, Cambridge MA

Nicoud FR (2004) The Structure of Simple 'New Economic Geography' Models. Working Paper No. 4326. Cente for Economic Policy Research (CEPR), London

Nijkamp P and Reggiani A (1998) The Economics of Complex Spatial Systems. Elsevier, Amsterdam

North DC (1993) Institutions and Credible Commitment. Journal of Institutional and Theoretical Economics (Zeitschrift für die gesamte Staatswissenschaft) 149: 11-23

Oerlemans LAG and Meeus MTH (2002) Spatial Embeddedness and Firm Performance: An Empirical Exploration of the Effects of Proximity on Innovative and Economic Performance. In: The 42nd Congress of the European Regional Science Association From Industry to Advanced Services. Perspectives of European Metropolitan Regions, Dortmund Germany

Oerlemans LAG, Meeus MTH and Boekema FWM (1998) Learning, Innovation and Proximity. An Empirical Exploration of Patterns of Learning: A Case Study. Working Paper No. 98.3. Eindhoven Centre for Innovation Studies, Eindhoven NL

Orlando MJ (2000) On the Importance of Geographic and Technological Proximity for R&D Spillovers: An Empirical Investigation. Working Paper No. RWP 00-02. Federal Reserve Bank of Kansas City, Kansas City

Orsenigo L (2001) The Failed Development of a Biotechnology Cluster: The Case of Lombardy. Small Business Economics 17: 77-92

Ottaviano GIP and Puga D (1997) Agglomeration in the Global Economy: A Survey of the 'New Economic Geography'. Working Paper No. 356. Centre for Economic Performance (CEPR), London

Ottaviano GIP and Thisse J-F (2002) Agglomeration and Trade Revisited. International Economic Review 43: 409-436

Overman HG, Redding S and Venables AJ (2001) The Economic Geography of Trade, Production and Income: A Survey of Empirics. Working Paper No. 2978. Centre for Economic Policy Research (CEPR), London

Palander T (1935) Beiträge zur Standorttheorie (Contributions to Location Theory). Almqvist and Wicksell, Uppsala Sweden

Paniccia I (1998) One, a Hundred, Thousands of Industrial Districts- Organizing Variety in Local Networks of Small and Medium-Sized Enterprises. Organization Studies 19: 667-699

Paniccia I (2002) Industrial Districts: Evolution and Competitiveness in Italian Firms. Edward Elgar, Cheltenham

Pavitt K (1987) On the Nature of Technology. University of Sussex, Science Policy Research Unit, Brighton

Perroux F (1950) Economic Space: Theory and Applications. Quarterly Journal of Economics 64: 89-104

Piore MJ and Sabel CF (1984) The Second Industrial Divide. Basic Books Inc., New York

Porter ME (1990) The Competitive Advantage of Nations. Macmillan, London

Porter ME (1998) The Competitive Advantage of Nations. 2nd Edition, Palgrave, New York

Porter ME (2000) Location, Competition and Economic Development: Local Clusters in a Global Economy. Economic Development Quarterly 14: 15-34

Powell WW (1998) Learning from Collaboration: Knowledge and Networks in the Biotechnology and Pharmaceutical Industries. California Management Review 40: 228-240

Powell WW and DiMaggio PJ (1991) The New Institutionalism in Organizational Analysis. The University of Chicago Press, Chicago

Powell WW, Koput KW and Smith-Doerr L (1996) Interorganizational Collaboration and the Locus of Innovation: Networks of Learning in Biotechnology. Administrative Science Quarterly 41: 116-145

Prigogine I (1997) The End of Certainty: Time, Chaos, and the New Laws of Nature. The Free Press, New York

Prigogine I and Stengers I (1984) Order out of Chaos: Man's New Dialogue with Nature. Bantam Books, Toronto

Puga D (1998) The Rise and Fall of Regional Inequalities. European Economic Review 43: 303-334

Puga D and Venables AJ (1996) The Spread of Industry: Spatial Agglomeration in Economic Development. Working Paper No. 1354. Centre for Economic Performance (CEPR), London

Pumain D (2004) Scaling Laws and Urban Systems. Working Paper No. 04-02-002. Santa Fe Institute, Santa Fe, New Mexico

Pyke F, Beccattini G and Sengenberger W (1990) Industrial Districts and Inter-Firm Co-Operation in Italy. International Institute for Labour Studies, Geneva

Pyke F and Sengenberger W (1990) Introduction. In: Pyke F, Beccattini G and Sengenberger W (eds) Industrial Districts and Inter-Firm Co-Operation in Italy. International Institute for Labour Studies, Geneva, pp 1-9

Quah D (2002) Spatial Agglomeration Dynamics. American Economic Review 92: 247-252

Quelle O (1926) Industriegeographie der Rheinlande (The Industrial Geography of the Rhineland). Schroeder, Bonn

Quevit M (1991) Innovative Environments and Local/ International Linkages in Enterprise Stratey: A Framework for Analysis. In: Camagni R (ed) Innovation Networks. Belraven Press, London, pp n.a.

Rabellotti R (1997) The Industrial District Model. In: Rabellotti R (ed) External Economies and Cooperation in Industrial Districts - a Comparison of Italy and Mexico. St. Martin's Press, New York, pp 23-43

Rabellotti R (1995) Is There an Industrial District Model? Footwear Districts in Italy and Mexico Compared. World Development 23: 29-41

Ratti R, Bramanti A and Gordon R (1997) The Dynamics of Innovative Regions - The GREMI Approach. Ashgate, Aldershot UK

Rauch JE (1993) Does History Matter Only When It Matters Little? The Case of City-Industry Location. Working Paper No. 4312. National Bureau of Economic Research (NBER), Cambridge, MA

Richardson HW (1993) Economies and Diseconomies of Agglomeration. In: Giersch H (ed) Urban Agglomeration and Economic Growth. Springer, Heidelberg, pp 123-156

Ritschl H (1927) Reine und historische Dynamik des Standortes der Erzeugungszweige ('Pure' and Historic Dynamics of the Location of Manufacturing Industries). Schmollers Jahrbuch 51: 813-870

Rivkin JW (2000) Imitation of Complex Strategies. Management Science 46: 824-844

Rivkin JW and Siggelkow N (2003) Balancing Search and Stability: Interdependencies among Elements of Organizational Design. Management Science 49: 290-311

Robert-Nicoud F (2004) The Structure of Simple 'New Economic Geography' Models. Working Paper No. 4326. Centre for Economic Policy Research (CEPR), London

Roos M (2002a) How Important Is Geography for Agglomeration? In: From Industry to Advanced Services - Perspectives of European Metropolitan Regions, The European Regional Science Association (ERSA), Dortmund, Germany

Roos M (2002b) Ökonomische Agglomerationstheorien: Die Neue Ökonomische Geographie im Kontext (Economic Agglomeration Theories: The Context of the New Econonmic Geography). Josef Eul Verlag, Lohmar Germany

Rosa P and Scott M (1999) Entrepreneurial Diversification, Business-Cluster Formation, and Growth. Environment and Planning C: Government and Policy 17: 527-547

Rosenberg D (2002) Cloning Silicon Valley. The Next Generation of High-Tech Hotspots. Pearson, Edinburgh

Rosenthal SS and Strange WC (2001) The Determinants of Agglomeration. Journal of Urban Economics 50: 191-229

Rosenthal SS and Strange WC (2003) Geography, Industrial Organization, and Agglomeration. Review of Economics and Statistics 85: 377-393

Ross EA (1896) The Location of Industries. Quarterly Journal of Economics 10: 247-268

Rullani E (2002) The Industrial Cluster as a Complex Adaptive System. In: Curzio AQ and Fortis M (eds) Complexity and Industrial Clusters: Dynamics and Models in Theory and Practice. Physica, Heidelberg, pp 35-64

Russo M (1985) Technological Change and the Industrial District. Research Policy 14: 329-343

Samuelson PA (1954) The Transfer Problem and Transport Costs, II: Analysis of Trade Impediments. Economic Journal 64: 264-289

Sax E (1918-1922) Die Verkehrsmittel in Volks- und Staatswirtschaft, Band I (Transportation in the National and State Economy, Volume I). Hölder, Wien

Saxenian A (1991) The Origins and Dynamics of Production Networks in Silicon Valley. Research Policy 20: 423-437

Saxenian A (1994) Regional Advantage: Culture and Competition in Silicon Valley and Route 128. Harvard University Press, Cambridge MA

Schamp EW (2000) Decline and Renewal in Industrial Districts: Exit Strategies of SMEs in Consumer Goods Industrial Districts of Germany. In: Taylor M and Vatne E (eds) The Networked Firm in a Global World: Small Firms in New Environments. Aldershot, Ashgate UK, pp 257-281

Schelling TC (1978) Micromotives and Macrobehavior. W. W. Norton & Company, New York

Schlier O (1922) Der deutsche Industriekörper seit 1860: Allgemeine Lagerung der Industrie und Industriebezirksbildung (The Body of the German Industry since 1869: General Industry Distribution and the Emergence of Industry Districts). Mohr, Tübingen Germany

Schmitz H (1999) Responding to Global Pressure: The Role of Private Partnership and Public Agencies in the Sinos Valley, Brazil. In: Inter-American Development Bank Conference: Building a Modern and Effective Business Development Services Industry in Latin America and the Caribbean, Rio de Janeiro, Brazil

Schmutzler A (1999) The New Economic Geography. Journal of Economic Surveys 13: 355-379

Schoenberger E (1997) The Cultural Crisis of the Firm. Blackwell, Cambridge MA

Schumacher (1910) Die Wanderungen der Großindustrie in Deutschland und in den Vereinigten Staaten (The Dynamics of Large Scale Industries in Germany and the United States). Schmollers Jahrbuch 43: 451-481

Schumpeter JA (1934a) The Fundamental Problem of Economic Development. In: Schumpeter JA (ed) The Theory of Economic Development: An Inquiry into Profits, Capital, Interest and the Business Cycle. Harvard University Press, Cambridge MA, pp 57-94

Schumpeter JA (1934b) The Theory of Economic Development: An Inquiry into Profits, Capital, Interest and the Business Cycle. Harvard University Press, Cambridge MA

Schweizer F (1997) Self-Organization of Complex Structures: From Individual to Collective Dynamics. Gordon and Breach Science Publishers, Amsterdam

Scitovski T (1954) Two Concepts of External Economies. Journal of Political Economy 62: 143-151

Scott AJ (1988) New Industrial Spaces: Flexible Production and Regional Development in North America and Western Europe. Pion, London

Scott AJ (1998) Regions and the World Economy: The Coming Shape of Global Production, Competition, and Political Order. Oxford University Press, Oxford

Scott AJ (2004) A Perspective of Economic Geography. Journal of Economic Geography 4: 479-499

Sforzi F (1990) The Quantitative Importance of Marshallian Industrial Districts in the Italian Economy. In: Pyke F, Beccattini G and Sengenberger W (eds) Industrial Districts and Inter-Firm Co-Operation in Italy. International Institute for Labour Studies, Geneva, pp 75-107

Shefer D and Frenkel A (1998) Local Milieu and Innovations: Some Empirical Results. The Annals of Regional Science 32: 185-200

Siggelkow N and Levinthal DA (2003) Temporarily Divide to Conquer: Centralized, Decentralized, and Reintegrated Organizational Approaches to Exploration and Adaptation. Organization Science 14: 650-669

Simon HA (1955) A Behavioural Model of Rational Choice. Quarterly Journal of Economics 69: 99-118

Simon HA (1981) The Sciences of the Artificial. MIT Press, Cambridge, MA

Simon HA (2002) Near Decomposability and the Speed of Evolution. Industrial and Corporate Change 11: 587-599

Skyttner L (1996) General Systems Theory - an Introduction. Macmillan Press Ltd, Basingstoke HA

Sorenson O (2003) Social Networks and Industrial Geography. Journal of Evolutionary Economics 13: 513-527

Soubeyran A and Weber S (2002) District Formation and Local Social Capital: A (Tacit) Co-Opetition Approach. Journal of Urban Economics 52: 65-92

Squazzoni F and Boero R (2002) Economic Performance, Inter-Firm Relations and Local Institutional Engineering in a Computational Prototype of Industrial Districts. Journal of Artificial Societies and Social Simulation 5: n.a.

Staber UH (1996a) Accounting for Variations in the Performance of Industrial Districts: The Case of Baden Württemberg. International Journal of Urban and Regional Research 299-316

Staber UH (1996b) The Social Embeddedness of Industrial District Networks. In: Staber UH (ed) Business Networks: Prospects for Regional Development. de Gruyter, Berlin, pp 148-174

Steinle C and Schiele H (2002) When Do Industries Cluster? A Proposal on How to Assess an Industry's Propensity to Concentrate at a Single Region or Nation. Research Policy 31: 849-858

Storper M (1995) The Resurgence of Territorial Economies, Ten Years Later: The Region as a Nexus of Untraded Interdependencies. European Urban and Regional Studies 2: 191-221

Storper M (1997) The Regional World - Territorial Development in a Global Economy. The Guilford Press, New York

Storper M and Scott AJ (1988) The Geographical Foundations and Social Regulation of Flexible Production Complexes. In: Wolch J and Dear M (eds) The Power of Geography: How Territory Shapes Social Life. Allen and Unwin, London, pp 21-40

Stuart T and Sorenson O (2003) The Geography of Opportunity: Spatial Heterogeneity in Founding Rates and the Performance of Biotechnology Firms. Research Policy 32: 229-253

Sturgeon TJ (2000) How Silicon Valley Came to Be. In: Kenney M (ed) Understanding Silicon Valley: The Anatomy of an Entrepreneurial Region. Stanford University Press, Stanford CA, pp 15-47

Swann P (1996) Technology Evolution and the Rise and Fall of Industrial Clusters. Revue Internationale de Systémique 10: 285-302

Swann P, Prevezer M and Stout D (1998) The Dynamics of Industrial Clustering: International Comparisons in Computing and Biotechnology. Oxford University Press, Oxford

Sweeney G (1995) National Innovation Policy or a Regional Innovation Culture. Working Paper No. 1. Working Papers in European Industrial Policy, The European Network on Industrial Policy (EUNIP), Vienna

Tiberi Vipraio P (1996) From Local to Global Networking - the Reconstruction of Italian Industrial Districts. Journal of Industry Studies 3: 135-153

Tichy G (2001) Regionale Kompetenzzyklen - Zur Bedeutung von Produktlebenszyklus- und Clusteransätzen im regionalen Kontext (Cycles of Regional Competence - on the Importance of Product Life Cycle and Cluster Theory in a Regional Context). Zeitschrift für Wirtschaftsgeographie 45: 181-201

Toffler A (1984) Foreword: Science and Change. In: Prigogine I and Stengers I (eds) Order out of Chaos: Man's New Dialogue with Nature. Bantam Books, Toronto, pp xi-xxvi

Townroe PM (1972) Some Behavioural Considerations in the Industrial Location Decision. Regional Studies 6: pp. 261-272

Urnaga MG, Extebarria G and Cooke P (1998) Regional Systems of Innovation: An Evolutionary Perspective. Environment and Planning A 30: 1563-1584

Uzzi B (1997a) Social Structure and Competition in Interfirm Networks: The Paradox of Embeddedness. Administrative Science Quarterly 42: 35-67

Uzzi B (1997b) Towards a Network Perspective on Organizational Decline. The International Journal of Sociology and Social Policy 17: 111-155

Valente M (2005) Qualitative Simulation Modelling. In: 4th European Meeting on Applied Evolutionary Economics (EMAEE), Utrecht NL

van Dijk MP (1995) Flexible Specialisation, the New Competition and Industrial Districts. Small Business Economics 15-27

van Dijk MP and Pellenbarg PH (1999) Demography of Firms. Spatial Dynamics of Firm Behaviour. Koninklijk Nederlands Aardrijkskundig Genootschap, Utrecht NL

van Tulder R (1988) Small European Countries in the International Telecommunications Struggle. In: Freeman C and Lundvall B-A (eds) Small Countries Facing the Technological Revolution. Pinter Publishers, London, pp 169-183

Varian HR (1990) Monitoring Agents with Other Agents. Journal of Institutional and Theoretical Economics 146: 153-174

Venables AJ (1996) Equilibrium Locations of Vertically Linked Industries. International Economic Review 37: 341-360

Venables AJ (1998) Agglomeration, Comparative Advantage, and International Specialization: A Multi-Industry Model. Working Paper No. 1961. Centre for Economic Policy Research (CEPR), London

Venables AJ (2001) Geography and International Inequalities: The Impact of New Technologies. Working Paper No. 05/07 (September 2001). London School of Economics, London

von Beckerath E (1918-1922) Wandelungen der Wirtschaft im Zeitalter der Eisenbahnen (Economic Change in the Railway Era). In: Sax E (ed) Die Verkehrsmittel in Volks- und Staatswirtschaft (Transportation in the National and State Economy, Volume I). Springer, Berlin, pp n.a.

von Thünen JH (1826) Der Isolierte Staat in Beziehung auf Landwirtschaft und National- ökonomie (The Isolated State in Relation to Agriculture and the National Economy). Fischer Publications, Jena

Washington M and Ventresca MJ (2004) How Organizations Change: The Role of Institutional Support Mechanisms in the Incorporation of Higher Education Visibility Strategies, 1874-1995. Organization Science 15: 82-97

Weber A (1909) Über den Standort von Industrien (Theory of the Location of Industries). Mohr, Tübingen

Weidlich W (2000) Sociodynamics - A Systematic Approach to Mathematical Modelling in the Social Sciences. Harwood Academic Publishers, Amsterdam

Wells R (2001) How Institutional Theory Speaks to Changes in Organizational Populations. Health Care Management Review 26: 80-84

Wheeler CH (2003) Evidence on Agglomeration Economies, Diseconomies, and Growth. Journal of Applied Econometrics 18: 79-104

Wicks D (2001) Institutionalized Mindsets of Invulnerability: Differentiated Institutional Fields and the Antecedents of Organizational Crisis. Organization Studies 22: 659-692

Williamson OE (1975) Markets and Hierarchies - Analysis and Anti-Trust Implications. Free Press, New York

Wilson AG (2000) Complex Spatial Systems - The Modelling Foundations of Urban and Regional Analysis. Prentice Hall, Harlow UK

Witt U (1997) 'Lock-in' Vs. 'Critical Masses' - Industrial Change under Network Externalities. International Journal of Industrial Organization 15: 753-773

Wolter K (2005) High-Tech Industry Clustering Rationales: The Case of German Biotechnology. In: Piccaluga A (eds) Regional Economies as Knowledge Laboratories. Edward Elgar, Cheltenham, pp 117-141

Wright S (1931) Evolution in Mendelian Populations. Genetics 16: 97-159

You J-L and Wilkinson F (1994) Competition and Co-Operation: Toward Understanding Industrial Districts. Review of Political Economy 6: 259-278

Zeitz G, Mittal V and McAulay B (1999) Distinguishing Adoption and Entrenchment of Management Practices: A Framework for Analysis. Organization Studies 20: 741-776

Zucker LG, Darby MR and Armstrong J (1998) Geographically Localized Knowledge: Spillovers or Markets? Economic Enquiry 36: 65-86

Zucker LG, Darby MR and Brewer MB (1994) Intellectual Capital and the Firm: The Technology of Geographically Localized Knowledge Spillovers. Working Paper No. 4946. National Bureau of Economic Research (NBER), Cambridge, MA

Zucker LG, Darby MR and Brewer MB (1997) Intellectual Human Capital and the Birth of U.S. Biotechnology Enterprises. American Economic Review 88: 290-306

Printing and Binding: Strauss GmbH, Mörlenbach